THEORY AND EXPERIMENT

SYNTHESE LIBRARY

STUDIES IN EPISTEMOLOGY,

LOGIC, METHODOLOGY, AND PHILOSOPHY OF SCIENCE

VOLUME 195

THEORY AND EXPERIMENT

Recent Insights and New Perspectives on Their Relation

EDITED BY

DIDERIK BATENS

Rijksuniversiteit Gent and Vrije Universiteit Brussel

and

JEAN PAUL VAN BENDEGEM

Bevoegdverklaard Navorser N.F.W.O.
Rijksuniversiteit Gent and Vrije Universiteit Brussel

on behalf of the Joint Commission of the Sixth International Conference on the History and Philosophy of Science

Diderik Batens (Gent-Brussel)
Paolo Galluzzi (Firenze)
Robert Halleux (Liège)
Erwin N. Hiebert (Harvard)
Risto Hilpinen (Turku)

D. REIDEL PUBLISHING COMPANY

A MEMBER OF THE KLUWER ACADEMIC PUBLISHERS GROUP

DORDRECHT / BOSTON / LANCASTER / TOKYO

Library of Congress Cataloging in Publication Data

CIP

International Conference on the History and Philosophy of Science (6th: 1986: Ghent, Belgium and Brussels, Belgium)

Theory and experiment: recent insights and new perspectives on their relation / edited by Diderik Batens and Jean Paul Van Bendegem on behalf of the joint commission of the Sixth International Conference on the History and Philosophy of Science.

p. cm. — (Synthese library ; v. 195)

Sponsored by the International Union of the History and Philosophy of Science, held in Ghent and Brussels, Aug. 25–30, 1986.

Includes index.

ISBN-13: 978-94-010-7794-1 e-ISBN-13: 978-94-009-2875-6

DOI: 10.1007/ 978-94-009-2875-6

1. Science—Methodology—Congresses. 2. Science—Experiments—Congresses. 3. Science—History—Congresses. 4. Science—Philosophy—Congresses. I. Batens, Diderik, 1944– . II. Bendegem, Jean Paul Van, 1953– . III. International Union of the History and Philosophy of Science. IV. Title. V. Series.

Q174.I563 1986

502'.8—dc 19

87–28870

CIP

Published by D. Reidel Publishing Company,
P.O. Box 17, 3300 AA Dordrecht, Holland.

Sold and distributed in the U.S.A. and Canada
by Kluwer Academic Publishers,
101 Philip Drive, Norwell, MA 02061, U.S.A.

In all other countries, sold and distributed
by Kluwer Academic Publishers Group,
P.O. Box 322, 3300 AH Dordrecht, Holland.

TABLE OF CONTENTS

PREFACE

This is not "another collection of contributions on a traditional subject." Even more than we dared to expect during the preparatory stages, the papers in this volume prove that our thinking about science has taken a new turn and has reached a new stage.

The progressive destruction of the received view has been a fascinating and healthy experience. At present, the period of destruction is over. A richer and more equilibrated analysis of a number of problems is possible and *is being carried out*. In this sense, this book comes right on time.

We owe a lot to the scholars of the Kuhnian period. They not only did away with obstacles, but in several respects instigated a shift in attention that changed history and philosophy of science in a irreversible way. A clearcut example - we borrow it from the paper by Risto Hilpinen - concerns the study of science as a process, and not only as a result. Moreover, they apparently reached several lasting results, e.g., concerning the tremendous impact of theoretical conceptions on empirical data. Apart from baffling people for several decades, this insight rules out another return to simple-minded empiricism in the future.

However, the present turn in our thinking about science is *also* essentially dependent on ideas of more recent vintage. A case at hand is the study of and the new insights in the generative role of experience in general and of experiment in particular. During the first decades of our century the problem was considered to be without any methodological interest. More recently, the emphasis on the precedence of theories put the problem largely out of focus or at least prevented a suitable 'representation'. At present, the generative role of experiments receives extensive attention and suspicions about the inquiry and its results have vanished. The subject is the main concern of Thomas Nickles' contribution to this book. It also is explicitly discussed in the two more general analytic papers, by which this volume opens, and is the main concern of several historically oriented papers.

Let us take one more example to illustrate the new approach. The role of experiments in *testing* theories is a traditional problem. It has been central both for the Logical-Empiricists and during the Kuhnian period. A theme that has been considered essential all the time is the Quine-Duhem thesis. For almost a century now, the attitudes towards it have been oscillating between a rejection which reasoned around the very difficulty, and an acceptance at the expense of answering justifiability problems.

The reader who is familiar with the story will appreciate Allan Franklin's approach to the problem.

Before we say a word about the papers, we want to make one more general point. As is well-known, science and philosophy are, for different reasons, under a rather heavy strain. A growing number of people feels the need for a new approach to nature and man, and as a consequence also to the epistemological relation. A central issue, in the social sciences as well as in mathematics, in epistemology as well as in social discussions, concerns the multiplicity of relations between normative and factual components of knowledge systems, belief systems, as well as social systems. Neither history of science nor philosophy of science is concerned with the general solution to this problem. Yet, in both disciplines has grown a practice in which normative and descriptive features are treated as basically belonging to the same domain. Especially during recent years this practice has become firmly established, even if the underlying methods are not sufficiently spelled out or reflected upon. For this reason, volumes as the present one form important materials for the development of the new approach to knowledge.

We have divided the papers in three sections. Given the firmly established interaction between historical and philosophical studies, we judged a division according to the two disciplines inappropriate. We realize that any non-external classification involves an interpretation, but trust that the reader will also value each separate paper for its own merits and with respect to its own aims.

The contributions by Evandro Agazzi and Risto Hilpinen analyze the general relations between theory and experiment. Stressing the importance of questions, both relate to the growingly popular problem-solving approach; Hilpinen's analysis explicitly proceeds within a question-answering framework that aims at a formal basis for this approach. Agazzi highlights the distinction between genetical and logical questions, Hilpinen the distinction between primary and secondary questions. In doing so, they are able to clearly phrase the role of presuppositions, background knowledge, and topics related to consistency and error.

The second part of the book contains papers on the two major roles of experiments with respect to theory: generation and testing. Thomas Nickles' paper continues the analysis of the two previous papers, but concentrates on a general overview of the aforementioned roles. As the reader will expect, he offers extensive arguments for the generative role of experiments, thus further substantiating the recently developed tenets on the logic of discovery. This role of experiments is central in the case studies presented by Erwin Hiebert and Christoph Meinel. Hiebert shows that, during the early thirties, experimental research on the structure of the atom led a relatively independent life, while theorists were struggling desperately to get a grip on its results. Meinel's detailed study of seventeenth-century atomistic views, partly aiming at a rectification of the historical record, emphasizes that the rationale of the intriguing experiments was the formulation of theories. Robert Halleux also wants to qualify the status of some of the 'experiments' of that period. His study of van Helmont's work results in a taxonomy which is acceptable by present-day standards but enables us to understand the confusion of those days. The contributions by Nils-Eric Sahlin, Edward Erwin, and Allan Franklin all concentrate on the function of experiments in *testing* theories. They critically examine the reliability of experimental results and their suitability for the evaluation, rejection, or acceptance of theories. The papers concentrate on a variety of applications: Freud's theory, *normative* decision theory, more particularly Bayesianism, and theories within the field of elementary particle physics. In each paper, a set of methodological distinctions, rules, and maxims is argued for with respect to the particular case study, but is intended as being of general application.

The third part of the book contains the papers that concern the role of theoretical conceptions. They deal, in general or with respect to specific cases, with such features as 'focus', 'theory-ladenness', and 'wholeness'. In her study of Heisenberg, in which the importance of visualization is illustrated, Catherine Chevalley shows the influence of theoretical changes on the very notion of experiment. Gérard Simon analyses Ptolemy's experiments. He sets out to show that the experimental techniques of antiquity were not in any sense inferior to ours, but that the surrounding conceptual framework resulted in interpretations which cannot be related to any present-day theory. Huldrych M. Koelbing offers a mainly conceptual case study of Newton's and Goethe's color theories. He describes the history of the conceptual systems of both, as well as their similarities and differences. Starting from a qualitative-network view, Aristides Baltas argues that the most important experiments are selected by the theory itself. He highlights the role of novel concepts in setting up experiments, and stresses the inevitability of breaking with intuitive notions. Reflecting on anthropological field-work and its theoretical foundation, Silvana Borutti studies the role of dialogues as a determinant in the definition of experience and report. Vincenzo Cappelletti relies largely on an analysis of the conceptions of Mach and Dilthey to defend a synthetic, dialectical view, as opposed to a binary view, on paradigm change. Marcello Pera's thesis is that traditional views of scientific rationality should be supplemented with argumentative components, if they are to offer an understanding of the history of science; he takes numerous illustrations from the writings of Darwin.

The papers presented in this volume derive from the "Sixth Joint International Conference on the History and Philosophy of Science". Some information about the conference is listed on the following pages. One participant informed us that he had promised his paper elsewhere. Some others were unable to meet the deadline for the book. Paul Weingartner was among the latter; a related paper of his, on the logical paradoxes and their methodological relevance, will appear in the proceedings of the 1986 Heidelberg meeting of the Académie Internationale de Philosophie des Sciences, edited by E. Scheibe and to be published by deGruyter, Berlin.

The present volume contains neither the paper on relevant logical results, nor the two papers related to the sociology of science. Apart from this, we believe it offers a good survey of both the diversity and unity we experienced at the conference.

Acknowledgments
The conference was sponsored by the International Union of History and Philosophy of Science. It obtained the moral support of the Koninklijke Academie voor Wetenschappen, Letteren en Schone Kunsten van België and of the Académie Royale des Sciences, des Lettres et des Beaux-Arts de Belgique. Financial support for the conference and the proceedings was obtained from the following authorities and agencies: International Union of History and Philosophy of Science, Nationaal Fonds voor Wetenschappelijk Onderzoek - Fonds National de la Recherche Scientifique, Ministerie van de Vlaamse Gemeenschap, Faculteit van de Letteren en Wijsbegeerte van de Rijksuniversiteit Gent, Centre National de Recherches de Logique - Nationaal Centrum voor Logica, Ministerie van Nationale Opvoeding, Ministère de l'Education Nationale. A small grant was obtained from Kredietbank. Some material help was provided by several departments of Rijksuniversiteit Gent.

The help of Mrs. Mady Van der Schelden in preparing this book has been invaluable. Nicholas Batens spent a long week-end helping us out with the final touch to the index. We want to thank the members of the organizing committee and of the joint commission for their advice and help, and most of all the participants in the conference for making it a valuable experience.

The editors

The "Sixth Joint International Conference on the History and Philosophy of Science" was organized under the auspices of the International Union of History and Philosophy of Science in Ghent and Brussels from 25 to 30 August 1986. The theme of the conference was: "Theory and Experiment".

The organizing committee was composed as follows: *Honorary Chairman*: Leo Apostel (Gent) - *Chairmen:* Diderik Batens (Gent-Brussel), Michel Meyer (Bruxelles-Mons) - *Secretary*: Jean Paul Van Bendegem (Gent-Brussel-Antwerpen) - *Other members*: Evandro Agazzi (Fribourg), P. Bockstaele (Leuven), Paul Gochet (Liège), P.V. Grosjean (Mons), Rudolf Haller (Graz), Robert Halleux (Liège), G. Hirsch (Bruxelles), Jean Ladrière (Louvain), Larry Laudan (Blacksburg), Herman Roelants (Leuven), J. Ruytinx (Brussel), Fernand Vandamme (Gent), Etienne Vermeersch (Gent), J. Vuillemin (Collège de France).

The International Union of History and Philosophy of Science had formed the following Joint Commission: Diderik Batens (Gent-Brussel), Paolo Galluzzi (Firenze), Robert Halleux (Liège), Erwin N. Hiebert (Harvard), Risto Hilpinen (Turku).

PROGRAM

MONDAY, AUGUST 25

Diderik Batens, *Welcome Address.*
Evandro Agazzi, *Do experiments depend on theories or theories on experiments ?*

Allan Franklin, *Experiment, theory choice, and the Duhem-Quine problem*
Marcello Pera, *Breaking the link between methodology and rationality. A plea for rhetoric in scientific inquiry*

TUESDAY, AUGUST 26

G. Simon, *Since the experimental method dates from antiquity, what exactly happened in the XVIIth century ?*
Christoph Meinel, *Early seventeenth-century atomism: theory, epistemology, and the insufficiency of experiment*
Robert Halleux, *Theory and experiment in the early writings of Johan Baptist Van Helmont*

Huldrych M. Koelbing, *Newton's and Goethe's colour theories - contradictory or complementary approaches ?*
Catherine Chevalley, *Heisenberg's notion of what is a physical theory*

WEDNESDAY, AUGUST 27

Risto Hilpinen, *On experimental questions*
Paul Weingartner, *Solving paradoxes in the logical relations between theory and experiment*
Vincenzo Cappelletti, *On the dynamics of paradigms*

THURSDAY, AUGUST 28

Edward Erwin, *Testing Freudian Hypotheses*
Silvana Borutti, *Models and interpretation in human sciences*
Nils-Eric Sahlin, *The significance of empirical evidence for recent developments of the foundations of decision theory*

Victor Tishchenko, *Social experiment and biological theory.*
Imre Hronszky, *Theory and experience: the knowledge-sociologist's intervention.*

FRIDAY, AUGUST 29

Thomas Nickles, *Reconstructing science: discovery and experiment*
William Harper, *Newton's inferences from the phenomena in the Principia*
Erwin N. Hiebert, *The role of experiment and theory in the discovery of nuclear fission*

Aristides Baltas, *On a non-empiricist approach to the structure of physics.*
Peter Galison, *Traditions of instruments, traditions of arguments.*

SATURDAY, AUGUST 30

Karin Chemla, *What could be experimentation in the history of mathematics?*
M.L.G. Redhead, *The role of models in relating theory and experiment.*
Michel Meyer, *Goodbye address*

I

SYSTEMATIC ANALYSES

EVANDRO AGAZZI

DO EXPERIMENTS DEPEND ON THEORIES OR THEORIES ON EXPERIMENTS?

SOME STANDARD VIEWS

The constitution of every scientific discipline seems to follow a well known pattern, which was first outlined in the speculations of the founders of modern science, such as Bacon and Galilei, and which has undergone many refinements up to the philosophy of science of our century. According to this pattern, scientific inquiry begins with a systematic collection of empirical evidence, which we may call the level of *experience*, followed by the elaboration of a set of hypotheses, the level of *theories*, which are then submitted to test by means of *experiments*. Once these three aspects have reached a stage of mutual logical consistency, we usually say that a satisfactory "scientific theory" is available concerning our domain of investigation. While disagreement still exists regarding the transition from experience to hypotheses (e.g. some maintain that it occurs by induction, while others claim that it depends on bold conjectures), a more general agreement exists regarding the second point: experiments are designed as possible logical consequences of the hypotheses constituting the theory and they are able either to validate or invalidate those hypotheses. Disagreements emerge again regarding the third step, as a variety of positions exists concerning whether the negative outcome of experiments really entails the rejection of the theory. It is this last disagreement in particular that invites us to analyze the mutual dependence of experiments and theories.

Before continuing, it may be advisable to remark that throughout this paper we are going to use the notion of "experiment" in the rather technical and restricted sense outlined above, i.e. in the sense of an empirical procedure performed with a view to *testing* something, be it a tentative conjecture, a rather elaborated hypothesis, or perhaps a whole theory. Therefore we shall not use this notion to denote the simple observation of facts, even when it is systematic and rigorous, since this observation is specifically intended only to *ascertain* data. In other words, while pure observations may be characterized as attempts to know *how* certain processes transpire, experiments may be characterized as attempts

D. Batens and J. P. van Bendegem (eds.), Theory and Experiment, 3–13.

to know *whether* such processes transpire in a certain way or not. This distinction does not affect such more generic ways of speaking as are manifest in the expressions "experimental research," "experimental science," "experimental method,"and so on. Indeed, people doing experimental research may well be engaged in observations rather than in experiments (in the above sense), since in this case the word "experimental" is used rather to distinguish this activity from "theoretical" research. Similarly, the "experimental method" is a complex structure in which observations, hypotheses and experimental tests find their proper place, so it cannot simply be reduced to the moment in which experiments are performed. The same must be said of the "experimental sciences," which are those in which the experimental method is applied. In conclusion, we intend to use the notion of experiment in its most exact and restricted sense simply for the sake of our analysis, and have no objection to its more liberal use in other contexts.

TWO KINDS OF DEPENDENCE BETWEEN EXPERIMENTS AND THEORIES

According to the traditional view, experiments are designed to put theories to the test, and "theories" deserve this honourable denomination only if the test is successful. This seems to indicate a twofold dependence between experiments and theories. On the one hand we must say that experiments depend *genetically* and *logically* on theories, since they are conceived and designed as explicit questions formulated by means of the concepts of a theory and with a view to testing it. On the other hand we must say that a theory depends on experiments as far as its *legitimacy* is concerned, for the negative result of a relevant experiment would imply the elimination of the theory in question, thereby affecting its very *existence*.

On this line of reasoning we find a certain ambivalence in the notion of theory, which is understood first as meaning generically what is conceptually elaborated beyond the level of experience and has rather the character of a hypothetical construction (in keeping with the well known distinction between "observational" and "theoretical" concepts). But then it is also understood as a full fledged intellectual picture of reality, which can be accepted only if it accords well with reality. If we recognize and accept this bifurcation of meaning, we can say that experiments depend on theories in the first sense, while theories in the second sense depend on experiments. In this way our investigation might be closed with a rather obvious but trivial result. Moreover, this result would seem to rely upon the claim that theories in the most appropriate and strongest sense are final and unshakable pictures of reality, a view which enjoys very little support nowadays. Hence we shall approach our problem without such a claim, but at the same time take into account the distinctive roles of theories which are at the root of it.

THE CONSTRUCTION OF THEORIES

Let us begin by suggesting some distinctions regarding the specific roles of theories and experiments in science. It is appropriate to note that a theory develops out of a particular "point of view" which has led to the isolation of certain aspects of reality, aspects that are *intended* to be the object of the investigation. This means that these aspects have already been singled out by means of concepts which enable us to state *what* we intend to study, and these concepts have to be equipped with suitable operational criteria by means of which we can *refer* to reality under the specific point of view adopted. The combined presence of these concepts and operational criteria amounts to the determination of the "domain of objects" of the inquiry, and at the same time provides us with an initial quantity of empirical evidence concerning these objects, i.e. with a good deal of *data* given by the immediate application of the said criteria to reality. This is what we called the level of *experience* above.

We may also call it the level of *observation*, provided we do not consider observation to be a kind of passive "looking at things"; it is rather a kind of "looking into things" in the sense that it is performed within the framework of a given "point of view" and involves the concrete operational manipulation of physical tools. Theory emerges as an attempt to describe and understand the domain of objects so identified, by utilizing all the available information, which is concretely constituted by the previously collected data. This attempt cannot avoid being one of interpreting and explaining the data, which implies that it has two aspects:the elaboration of concepts and the formulation of statements by means of these concepts. (These two aspects however should not be considered as successive "steps," for concepts are always elaborated in mutual connection with other concepts, and this gives rise to statements in which their meanings take shape.) Along these lines, the construction of a theory may be seen as an expansion of the concepts which initially expressed the specific "point of view" on reality, an expansion which depends in part on the implicit potentialities of these concepts and in part on the information contained in the empirical statements describing the available data. All this amounts to the creation of new concepts and new statements, which deserve to be called "theoretical" not because they are opposed to "observational" concepts and statements (as was supposed to be the case according to a well known but untenable dichotomy), but because they are the product of theory-making and exist for the sake of "theorizing," i.e. for the sake of understanding and explaining the data. By way of summing up, we could say that the purpose of the theory is to provide a "conceptual space" - a frame of rationality or a condition of intelligibility - which accords with the level of *meaning and intension*, but is still in need of being connected with a *reference*.[1]

THE REFERENTIAL TASK OF EXPERIMENTS

The task of providing a reference for the meanings expressed in the sentences of a theory is the specific responsibility of experiments. This is why any relevant experiment is conceived and designed under the strict supervision of the theory, of which it should express a *referential logical consequence*. Its being a logical consequence of the theory makes it dependent upon it: but its being referential makes it dependent upon the operational criteria mentioned above and, in this respect, independent of the theory. It is of course true, as we have stressed, that these criteria themselves depend on the specific concepts which constitute the particular "point of view" adopted. But *these* concepts do not belong to the theory: they are common prerequisites both of the theory and of the operational criteria, and work as guidelines for the determination of the "domain of objects" of the discipline or theory envisaged. Once this is clear we can see that if the experimental test is successful, it provides the theory with a *guarantee of reference* (at least to a certain extent); if it is not successful, it indicates that the theory has *no reference* (at least to a certain extent).

Some refinements are needed concerning the referential purport of experiments. The most immediate, but misleading, idea associated with this claim could be that an experiment puts us into direct contact with the referents of the theory, in the sense that it enables us to observe them. Such a case is not totally excluded in the preliminary steps in which very simple conjectures of very elementary empirical theories are put to the test. However, such instances are rather exceptional and not very interesting. In the case of all mature natural sciences, experiments do not allow us to observe directly the intended referents of a theory,i.e. those objects which the theory is "speaking about" (such as, e.g., elementary particles). Here the experiments simply provide us with what we have intentionally called a "guarantee of reference," which is usually indirect, but reliable.

A very familiar example may help clarify the issue: a photograph which I see is for me a reliable guarantee of reference concerning the existence of the person or building of which it is a photo, although I do not actually see them, but simply their photo. Why is this? Because in normal circumstances no photo can be taken of something which does not exist and, moreover, a photo normally reproduces a good deal of optically detectable features of an object. Of course, the possibility exists that "false" photos be forged on purpose, or that a "true" photo be taken of an object under so peculiar conditions that it will give a misleading image of it. However, a careful inspection of the photo and the circumstances in which it was taken, and especially a comparison of it with other photos taken by other persons in different circumstances, and perhaps consideration of information coming from other sources (such as letters or records) - all this may increase the strength of the "guarantee of referentiality" up to the point of making it a practical certainty.

It is clear even from this elementary example that a not negligible number of "presuppositions" are involved in every referential claim which goes beyond the mere witnessing of the immediate evidence. Some of them concern matters of fact, others are fragments of causal explanations, and others are sketches of "theories" in a broad sense. In the case of the exact sciences the situation is analogous to that of everyday life, with some complications and some simplifications. The complications depend on the fact that the search for referential guarantees (i.e. the design of experiments) involves a very sophisticated theoretical, mathematical, and technological machinery, together with all the skills related to it. The simplification consists in the fact that only a few aspects of reality are involved and, therefore, the direction in which the research has to be done is better determined and the margins of possible equivocation are more restricted. After these refinements it should be clear in which sense it is appropriate to say that experiments have a referential purport: they show those empirical features which we expect as consequences of certain properties of our referents, and which we can understand only as such consequences. But more on this will appear in the sequel.

THE FEEDBACK FROM EXPERIMENTS ON THEORIES

It is important to stress that, each experiment being by necessity a very limited logical consequence of the theory,its guarantee or disproof of reference cannot concern but a restricted part of the theory. Or, in other words: a positive result ensures that at least certain parts of the theory are endowed with reference; many positive results ensure that several parts of the theory are endowed with reference,but it remains uncertain whether the whole theory enjoys such a guarantee. Symmetrically, one must say that one or more negative results prove that some aspects of the theory lack reference, but cannot exclude that other aspects have it. This way of looking at things seems to account in a more satisfactory manner for the well known facts that a finite number of "confirmations" are not sufficient for calling a theory true, while neither are a finite number of "falsifications" really sufficient for declaring it false. These facts are usually explained by considering scientific theories as equivalent to a logical conjunction of propositions, a point of view which has proved unsatisfactory in several respects. According to the approach advocated here, a theory may rather be seen as a more or less faithful representation of its intended referents: it can never be perfectly faithful, but this does not imply its rejection, so long as it is considered to be sufficiently faithful. Experiments exist precisely in order to make such a decision possible. But in order to see this clearly we need to develop some other considerations, which will lead us to a view of scientific theories much more different from the traditional view than we have been able to express thus far.

OTHER ASPECTS OF THE INTERDEPENDENCE OF THEORIES AND EXPERIMENTS

Following upon the above clarifications, we may try to characterize the role of theory and experiment by means of certain other distinctions, for example by saying that theories describe "possible worlds" in which the known phenomena (i.e. the data) may occur, while experiments try to single out the "real world" to which they belong This they do by ideally discarding all of those possible worlds which, though having a meaning, do not have a reference within the *intended domain of objects*. If we consider things this way we can better appreciate the fact that theories are always intrinsically hypothetical. This feature is usually interpreted as expressing the fact that we can never be *certain* of their truth; but it seems more adequate to recognize that theories' being hypothetical means that they express *possibilities*, and are intrinsically unable to do more than that because the task of providing a transition from possibility to reality belongs to something else, i.e. to experiments.

Another way of expressing these ideas is to say that theories can only determine the formulation of *questions*, but cannot provide the *answers* to these questions, which are expected to come from the experiments. By saying this it is clear and beyond doubt that the answer depends on the question, but only in so far as its *meaning* is concerned (this meaning also providing an indication of how to determine the answer). Whether the answer be positive or negative however does not depend on the question, but must rely upon other sources. In our case these are constituted by the operational criteria which, in performing the experiments, bring us into contact with the *intended referents* of the theory. Using a terminology borrowed from classical philosophy, we might say that theories determine the constitution of the *semantic logos* (i.e. the realm of concepts and intensions), while experiments are oriented to the construction of the *apophantic logos* (i.e. the realm in which something is affirmed or denied of some subject matter, which constitutes the reference or the extension of the discourse).

THE QUESTION OF THEORY-LADENNESS

We are afraid that the whole of what has been said thus far might induce an exaggerated estimation of the dependence of experiments on theories, which is likely to be conveyed by the fashionable doctrine of the theory-ladenness of all scientific sentences. This is why we want now to restrict the purport of the statement that the theory provides the meaning of the question which the experiment tries to answer. This statement is to be understood as follows: the theory provides the *significance* of the question, its *point*, its purpose and reason, as well as providing the question with some global meaning resulting from its position in the general context of the theory. But it is not as though the question would be *meaningless*, i.e. *not understandable* without the theory. This cannot be the case simply because, in order to be performed, an experimental test must be entirely describable in terms of operational concepts and their meanings, which, as we have already remarked, do not depend on the theory. This is why we need the theory in order arrive at the experiment, in order to design it (hence it depends "genetically" on the theory, as we have already said), but not in order to justify it. Once the experiment has been performed, it assumes an independent existence and is in no need of help, simply because it has the same character as the data which are the indisputable base which a theory is challenged to account for and which it cannot modify or dispense with.[2]

A HERMENEUTIC APPROACH TO THEORIES

We can make our discourse more understandable if we say that the doctrine of theory-ladenness fails to distinguish between the hermeneutical, the semantical, and the referential aspects of a theory. The traditional empiricist view of theories conceived of them as sets of propositions connected through logical links, and it reduced scientific explanation to formal deduction.

This view was to a large extent shared also by Popper, whose main difference from the empiricists in this regard was that he did not consider induction, but creative conjecturing, as the path leading to the establishment of the general hypotheses from which scientific explanations take their start. Both approaches had in mind the discovery of *single* hypotheses (which were confused with general laws) and considered theories as simply being the *logical conjunction* of these hypotheses. A serious shortcoming of this approach is its inability to account for theory change; and it is well known that a *semantic* integration was introduced, which stressed that the *meaning* of every concept is determined by its links with all the other concepts in the theory. The consequence of this was that there should be no way to compare theories on the basis of experimental tests, for one and the same test would not be considered the same within two different theories, since it would have a different meaning in each of them owing to the "theory-ladenness" of all the concepts involved.

This conclusion stands in conflict with the fact that theories are actually compared and sometimes also rejected, in spite of what is said by the supporters of the "incommensurability" thesis. This indicates that these authors were right in stressing the global nature of theories, and in seeing that pure formal-logical conjunction is inadequate for expressing it, but that they were wrong in believing that this "wholeness" is of a semantical nature. In addition, they were unable to keep meaning distinct (though not totally separated) from reference.

Our claim is that the "wholeness" of a theory has to do with a *hermeneutic* level, which is connected with the *semantic* and the *referential* levels in different ways. Since it would not be advisable to enter into too many details here, let us simply say that, when we start investigating reality from a certain "point of view", we already apply to it a particular *interpretation* (hermeneutic level), which guides our choice of concepts (semantic level) and our devising of operational testing procedures (referential level). This means that we begin our inquiry by referring to things and "seeing them in a certain manner," i.e. having in mind a certain *Gestalt*, or image, or model of them. Then we try to make this model explicit by expressing it by means of propositions; and these are the *hypotheses* which in their *unity* constitute the theory. However, since the model is always a "seeing how" (which necessarily oversteps the crude evidence of "seeing that"), it contains elements which can be correctly conceived, and therefore gives rise to *meaningful* sentences (semantic level), but sentences which might not correspond to actual features of our objects (referential level). This is why we try to check these additional features by means of experiments.

THE HERMENEUTIC APPROACH TO EXPERIMENTS

Every time we uncover new facts which are in keeping with our *Gestalt*, or model, we strengthen and enrich it through further details, and this also happens when we check whether all the details of the model are correct, by performing experiments. Therefore the outcome of the experiments *in any case* implies a modification of the model: either in the sense that it is enriched (when the outcome is positive), or in the sense that it has to be altered (when the outcome is negative). The extent of an alteration depends on the *relevance* of the details disproved by the negative result; and, again, this relevance cannot be appreciated on the basis of either logical features (from a logical point of view every proposition is exclusively either true or false), or semantical features (even disproved sentences remain meaningful). An appreciation of this relevance is only possible on a *hermeneutic* basis, i.e. by considering to what extent the unexpected result challenges the whole picture, the whole *Gestalt* or model. It may be that the result simply implies some correction leaving the substance of the model untouched; or it may be that it demands a complete *reorientation* of the interpretation and, thus, the replacement of the old model by a new one. But this is possible because the experiment has a referential purport, which enables it to break the "hermeneutic circle", instead of remaining prisoner of it.[3]

This, by the way, is also the reason why experiments usually provide means for comparing theories: indeed theories are compared on the basis of *referents* and not *meanings* and it is quite possible (indeed it is very common) that we are able to refer to the same objects in spite of interpreting them in different ways (hermeneutic level) and of thinking of them through different concepts (semantic level).

WHY THEORIES DEPEND, AFTER ALL, ON EXPERIMENTS

Once these points are clear, we can understand in which sense it is possible and even necessary to claim that theories depend (in a certain sense) on experiments. This is simply a consequence of the fact that theories, although structurally descriptions of "possible worlds," are constructed with a view to being descriptions or pictures of the "real world," at least in the following weak sense. The possible world which a given theory describes must include the features of the domain of objects the theory is about, and this entails not only the empirically known features, but also those which are as yet unknown, but which should exist according to the model. But in order to fulfil this requirement, a theory, as we have seen, has to undergo certain tests of referentiality concerning these additional features, i.e. it has to submit itself to the judgment of experiments which, besides supporting or weakening its referentiality claims, have the immediate effect of increasing the amount of empirical data it is *obliged to account for*. This is the reason why, without pretending that a theory should be eliminated simply on being confronted with some negative experimental outcomes, we must say that neither can it be considered satisfactory so long as these difficulties exist, and that it is expected to undergo the necessary corrections in order possibly to overcome them. But this is also the reason why theories are perpetually subject to change as a result of experiments: indeed, since every experiment always brings in something new, its interpretation produces a further elaboration of the model, which again contains something overstepping the immediate evidence, and hence calls for further experiments, in an interplay which progresses indefinitely without ever being circular.

Université de Fribourg

NOTES

1. The above is a very brief presentation of my view concerning the making of scientific theories as systems of sentences

regarding specific "objects," a view which I have developed in several other publications. Besides my book *Temi e problemi di filosofia della fisica*, Abete, Roma, 1974², I mention the following articles: "Les critères sémantiques pour la constitution de l'objet scientifique," in *La sémantique dans les sciences*, Office International de Librairie, Bruxelles, 1978, pp. 13–29; "The Concept of Empirical Data: Proposal for an Intensional Semantics of Empirical Theories," in M. Przelecki et al. (eds.), *Formal Methods in the Methodology of Empirical Sciences*, Reidel, Dordrecht, 1976, pp. 153–157; and "Eine Deutung der wissenschaftlichen Objektivität", in *Allgemeine Zeitschrift für Philosophie*, 3 (1978), pp. 20–47.

2. A more detailed treatment of the distinction between semantical and referential levels and of the possibility of theory comparison based upon this distinction can be found in my paper, "Commensurability, Incommensurability, and Cumulativity in Scientific Knowledge," in *Erkenntnis* 22 (1985), pp. 51–77.
3. This hermeneutic approach to scientific theories has been advanced in my publications mentioned in note 1, and has also been developed in a number of studies stressing the "gestalt-view" of theories, of which I should mention at least C. Dilworth, *Scientific Progress*, Reidel, Dordrecht, 1986³. More details on the specific views presented in the last part of this paper may be found in two recent articles of mine: "Dimensions historiques de la science et de sa philosophie", in *Diogène* 132 (1985), pp. 61–83 (also translated in the English edition of this journal), and "La funzione del modello nella scienza", in *Il quadrante scolastico*, Trento, 31(1986), pp. 52–62.

RISTO HILPINEN

ON EXPERIMENTAL QUESTIONS[1]

I

According to R. G. Collingwood, "every statement that anybody ever makes is made in answer to a question".[2] Collingwood takes this to be an important characteristic of scientific thinking:

> In proportion as a man is thinking scientifically when he makes a statement, he knows that his statement is the answer to a question and knows what that question is. In proportion as he is thinking unscientifically he does not know these things.[3]

Philosophers have studied science both as a product and as a process. The investigation of "the language of science", the structure of theories, and the conditions of their justification has been mainly concerned with science as a product; this line of investigation dominated the philosophy of science of the 1930's and 1940's. When philosophers became interested in questions of discovery, theory change and scientific progress, scientific rationality, and the interplay of theory and observation (or experiment) in science, they began to look at science as activity, viz. as a *process of inquiry*. In this context questions about the values and aims of science are unavoidable, and it is natural to adopt Collingwood's general standpoint: the purpose of inquiry is to find satisfactory answers to the questions in which scientists are interested. The interrogative accounts of scientific inquiry developed in recent years are attempts to spell out the meaning of this general observation in detail.[4] In the present paper I shall suggest that questions are methodologically relevant in two basic ways: The purpose of an inquiry is to find satisfactory answers to certain questions, and these answers can be found by presenting questions to various sources of information. Experiments can be regarded as questions of the second kind.

D. Batens and J. P. van Bendegem (eds.), Theory and Experiment, 15–29.

II

I take the main "product" of investigative activity to be a *belief system*, that is, a system of accepted propositions. An inquiry is a process by which a belief system which is in some respect unsatisfactory is modified and transformed into a new system. If the inquiry is successful, the resulting system is less unsatisfactory than the original one.

What does it mean to say that a system is "satisfactory" or "unsatisfactory"? This depends of course on the objectives of belief formation. Many philosophers have assumed that apart from various practical objectives, the evaluation of belief systems depends on two principal cognitive requirements, viz.

(i) the requirement of informational completeness, and

(ii) the requirement of error avoidance.[5]

Some philosophers have spoken in this context about "epistemic utilities" or "epistemic values".[6] The first objective reflects the value of *information*, the second objective, error avoidance, that of *truth*.[7]

Requirements (i) and (ii) seem absolute conditions on belief systems, but in fact they should be regarded as relative to the questions or problems in which investigators are interested. This is easy to see in the case of information (informational completeness). Scientists are not interested in all kinds of information, but only in information relevant to the problems which they are trying to solve. The evaluation of the informativeness of a belief system is always relative to some set of problems or questions. The evaluation of the truthfulness or error-freedom of a belief system can be relativised to a set of questions in an analogous way. The main indication of error in a belief system is inconsistency. Because of the general fallibility of our beliefs (including scientific beliefs), it seems reasonable to assume that even the most rational belief system may contain some false beliefs. But if we include in a given belief system a proposition to the effect that some beliefs in the system are false, the system becomes inconsistent, and this kind of inconsistency is not likely to be removed by any process of inquiry.[8] Thus even the most reasonable human belief systems may be inconsistent, and therefore unsatisfactory from the standpoint of absolute error avoidance. (I am assuming here that human belief systems are not logically closed.)

But this kind of inconsistency and error need not affect the ability of the system to provide consistent and seemingly error-free answers to the questions in which an investigator is interested: even an inconsistent system may be apparently error-free *relative to a set of questions* QS in the sense that it does not give inconsistent answers to any question Q_i in *QS*. If a belief system is apparently error free with respect to a question and contains an answer to it, we can say that the answer is apparently error-free with respect to the question. (I say that a belief system B_a "gives an answer" to a question Q if and only if B_a contains a proposition which is a complete answer to Q; the concept of "complete answer to a question" is here taken as given.)

The error-freedom of a system of beliefs is relative to questions in another way as well. Above error-freedom was treated as a qualitative, yes-no concept, but in some cases it seems possible to speak about the *degree of error* of a proposition or a system of propositions, and make comparisons between different propositions in this respect. In special cases it may even be possible to measure the *distance* of a proposition from the truth. This concept of "degree of falsity" is an important ingredient in Popper's concept of truthlikeness (verisimilitude).[9] David Miller has shown, however, that if the verisimilitude of a theory is measured by the accuracy of the predictions derived from it, no theory which is false (in the absolute sense) can have a higher degree of verisimilitude than another false theory.[10] Miller's result is essentially dependent on the assumption that all conceivable predictions derived from a theory are equally relevant to its evaluation. The result is avoided if the concept of verisimilitude - or degree of error - is relativised to a set of questions or problems. Thus Miller's result shows that if the "degree of error" of a system of beliefs is interpreted in terms of the *accuracy* of the beliefs, the assessment of the truthlikeness of the system must be relativised to a set of questions.[11]

III

The questions which determine the "goodness" of a person's belief system may be termed his *primary* questions. I assume that the main function of a person's belief system is to provide error-free and otherwise satisfactory answers to his primary questions. We might say that an investigator's primary question is a question addressed by the investigator to himself, or to his belief system, or a question which some (external) questioner (or "client") presents to the investigator and which the latter should be able to answer. (These questions should be distinguished from the questions which the investigator presents to some external source of information: the latter questions will be called below *secondary* questions; see section IV.)

Under what conditions does an investigator's belief system provide a "satisfactory" answer to a (primary) question Q? According to the traditional view of the pragmatics of questions, a question is an expression of a desire to know something. For example, according to Charles Peirce, a genuine or serious interrogation involves

> first, a sense that we do not know something; second, a desire to know it; and third, an effort - implying a willingness to labor - for the sake of seeing how the truth may really be.[12]

Lennart Åqvist and Jaakko Hintikka have developed a sophisticated formal version of this classical view in which questions are treated as *epistemic imperatives*.[13] According to the Åqvist-Hintikka semantics, a simple yes-no question

(1) Is it the case that p?

can be analysed as

(2) Bring it about that I know that p or I know that not-p,

or formally,

(3) $ID_b(K_a p \vee K_a{-p})$,

where 'I' is an expression for the imperative mood, 'D_b' is an action operator which stands for "b brings it about that", and 'K_a' is the epistemic operator, read as "a knows that".

Jaakko Hintikka calls the proposition in the scope of the action operator the *desideratum* of the question.[14] An addressee's answer to a question is satisfactory if and only if it brings about the epistemic state (state of knowledge) expressed by the desideratum. If we regard an investigator's primary questions as questions directed by the investigator to himself or to his belief system, the system gives a satisfactory answer to a question Q only when the investigator knows some proposition p such that p is a complete answer to Q. According to the usual interpretation of the concept of knowledge (which I accept here), this is the case only if

(i) p is true, and

(ii) p is justified within the investigator's belief system.[15]

We may say that an answer p is *prima facie* satisfactory if the investigator is justified in claiming or believing that he knows that *p*. If a person believes that *p* (accepts *p*) and *p* is justified within his belief system, then the person is also justified in claiming that he knows that *p*. If a person's belief system contains a proposition *q* which is inconsistent with *p*, *q* defeats any potential justification which B_a may contain for *p*, and consequently *p* cannot be justified within B_a under such circumstances. Thus a person's belief system cannot give a satisfactory answer to a question *Q* unless it is apparently error-free with respect to that question. But a belief system can be unsatisfactory in other ways as well: it may fail to give any answer at all to some primary question *Q*, or it may give only an answer which is unjustified, not because the system contains conflicting information about the question, but because the answer is not supported by any relevant evidence within the system. In all these cases the system needs improvement, and an inquiry is called for. Thus we arrive at the following classification of the circumstances or "knowledge situations" which are apt to initiate an inquiry:

(a) For some primary question *Q*, the investigator's belief system B_a contains no proposition *p* such that *p* is a complete answer to *Q*.

(b) B_a contains some complete answer *p* to *Q*, but no such answer is justified within B_a.

The second case (b) may arise in two ways, viz.

(b1) Proposition *p* is unfounded, that is, not supported by any evidence within B_a, or

(b2) *p* is not justified because B_a is not apparently error-free with respect to *Q*, that is, B_a also contains another answer to *Q* which is inconsistent with *p*.

In the first case, (b1), an inquiry is necessary in order to determine whether *p* can be justified, that is, whether it is reasonable to believe that *p* is true; in the second case it is necessary in order to decide which (if any) of the conflicting answers is the correct answer.

The view that the evaluation of a person's belief system is relative to a set of questions or problems agrees with Laurens Laudan's conception of the problem-solving effectiveness of theories as the main criterion of scientific progress. But Laudan's conception of problem solving seems (at least superficially) quite different from that proposed above. Laudan has argued that questions of truth, corroboration, and confirmation are in this context largely irrelevant;[16] he characterises a solution to a problem as follows:

> Generally, any theory T can be regarded as having solved an empirical problem, if T functions (significantly) in any

schema of inference whose conclusion is the statement of the problem.[17]

This characterisation is incomplete, because it makes no reference to the epistemic requirements for acceptable problem solutions. (Laudan does not deny the existence of such requirements.) I have assumed above that a solution is satisfactory only if it is apparently error-free with respect to the problem, and thus capable of being regarded as a (*prima facie*) true solution, and also supported by the investigator's other beliefs (in other words, only if it is justified within his belief system).

IV

An inquiry is a process by which an investigator attempts to transform a belief system which is unsatisfactory in some of the ways mentioned above into a more satisfactory system. One of the ordinary meanings of the verb "inquire" is "to ask a question", and I assume that an inquiry (in the sense discussed here) always takes place through questioning. But the questions which constitute an inquiry must be distinguished from those which necessitate it (the primary questions); the former questions are not directed to the investigator's own belief system, but to some external source of information. An inquiry is necessary when the investigator's own belief system does not give satisfactory answers to his primary questions, and he needs "help" from external sources. Under these circumstances the investigator becomes the client of some external source of information. The source of information used may be the investigator's own senses, a measuring instrument, an experimental apparatus, or any other source which is external to the investigator's own belief system. To distinguish the questions addressed to external sources from the investigator's primary questions I shall call them *secondary* questions or *experimental* questions. The expression "secondary question" indicates that the function of these questions is to find satisfactory answers to the primary questions; the expression "experimental question" suggests that experiments constitute an important subclass of these questions. Following Bacon and Kant, we may regard experiments as questions addressed to nature.[18] To bring out the contrast between primary questions and experimental questions, the former might also be called *theoretical* questions. (However, in certain respects this expression would be misleading; I shall return to this point below.)

The distinction between primary and secondary questions is not new. For example, in his interrogative theory of inquiry Jaakko Hintikka makes a distinction between a conclusion *C* which "the Inquirer" is trying to derive from his background theory *T*, and the questions which the Inquirer can present to "Nature" (or "the Oracle") in order to find additional premises for the deduction of *C* from *T*. The question whether *C* is the case is the "big" initial question which the Inquirer is trying to answer by presenting a number of "small" questions to Nature; the "big" question is what I have called above a "primary" question, whereas the questions presented to Nature are secondary questions.[19]

The justification requirement explains why a belief system which is not error-free with respect to a certain question, that is, which contains two mutually incompatible answers to a certain question, cannot usually be revised simply by deleting one of the conflicting answers: the remaining answer must be well-justified to be satisfactory. For this reason it is often necessary to expand a belief system be means of secondary questions even when the system contains several answers to some question: the expansion is necessary for making a justified choice among the alternatives.

Like the primary questions, secondary questions are requests for knowledge or epistemic requests. A given source of information can give the investigator knowledge only about certain secondary questions (or kinds of secondary questions). For this reason it is normally not possible to present primary questions directly to an external source: obtaining satisfactory answers to the primary questions requires demonstrative and non-demonstrative reasoning from the investigator's background beliefs and from the answers received to the secondary questions.

The so-called examination questions provide a good example of this difficulty. The Åqvist-Hintikka type semantics outlined earlier seems to be inapplicable to examination questions, because the examiner already knows the answers to the questions he presents to the examinee; for this reason many philosophers have regarded examination questions as a counterexample to the view that questions are epistemic requests.[20] However, it is clear that the purpose of examination questions is to give the examiner (or questioner) new information: not about the subject-matter of the examination (which is presumably already familiar to the examiner), but about the student's knowledge of the subject-matter.[21] Thus examination questions should be regarded as secondary questions. For example, in the case of an examination in history, the primary question in which the teacher is interested is: "How much does the student know about history?" If the teacher does not already know this, he must seek information from an external source, and in many cases the best source of information is the student himself. However, the teacher cannot present his primary question directly to the student: the answer given by the student to that question would not be trustworthy. Thus the teacher must perform an experiment and try to find an answer to his primary question by means of suitable secondary questions, that is, by means of an examination or *test*. He performs the test by asking the student questions and by registering the answers. (We might say that the teacher is in this questioning situation feigning ignorance about history.) If examinations are understood in this way as secondary questions (as tests or experiments), the answers to these questions should be regarded as propositions of the following form: "The student replies that *p*".

An answer received from an external source may fail to qualify as knowledge unless it coheres (or is supported by) the other beliefs in the questioner's belief system. An unexpected or unsupported experimental result is apt to give rise to new questions which are necessary for ascertaining the validity of the result. If the results of an experiment (or in general, the answers received to some secondary question) do not appear trustworthy, the experiment gives rise to a new primary question which must be decided by means of new secondary questions.

The main purpose of secondary questions is to give the investigator trustworthy information about the primary question (or questions) in which he is interested. Consequently the informational value of a secondary question *Qe* in a given situation depends on the amount of information transmitted by *Qe* about a certain primary question *Q*. This information can be measured by the expected value of the (amount of) information transmitted by a complete answer to *Qe* about a complete answer to *Q*.[22]

Above I mentioned that the secondary questions, like the primary questions, are requests for knowledge. However, since the main purpose of secondary questions is to give information relevant to the investigator's primary questions, a secondary question *R* which adds to the investigator's belief system false beliefs may also give him enough correct information about the subject-matter of the primary question under consideration to fulfill its purpose, i.e., to help the investigator find a correct answer to a primary question or justify a certain answer to the primary question. Relatively vague knowledge-claims can sometimes be justified by "sharp" (or informative) but false beliefs which are reasonably close to the truth. This seems to be common in experimental science, where (correct) theoretical claims are often justified by fallible and inaccurate experimental results. In his paper 'Experiment, Theory Choice, and the Duhem-Quine Problem' Allan Franklin has argued that incorrect experimental outcomes need not result in incorrect (or unjustified) theory choices. For example, Franklin notes that "the fact that Millikan's value of *e*, the charge of the electron ... disagrees with the currently accepted value ... has not changed the support for charge quantization".[23] This possibility seems a good counter-example to the foundationalist view that knowledge can be justified only by propositions which are true and known to be true.

V

As was pointed out above, the need to expand a belief system by means of secondary questions may arise in three ways: (a) B_a contains no (complete) answer to some primary question *Q*; (b1) B_a contains two or more conflicting answers to some question, or (b2) B_a contains an answer which is not justified within the system. In the first case, the purpose of the experimental question is to *find* (or "discover") an answer to the question *Q*, in the second case (b1) it is to *identify* the correct answer to the question from two or several alternatives, and in the third case (b2) the experimental questions serve the purpose of *testing* or *justifying* a putative answer. We may say that in a situation of type (a), the function of experimental questions is to increase the informativeness of the investigator's belief system with respect to a given primary question, and in situations (b), experimental questions are needed for improving the reliability of the system. The cases in which an investigator is, at least in his own opinion, fully justified in accepting a certain answer to a question, but his colleagues remain skeptical about the correctness of the answer, can also be regarded as instances of (b2). In this case the purpose of the investigator's secondary questions is to justify an answer for his colleagues rather than for himself.

In a recent paper Allan Franklin has discussed the question what makes a 'good' experiment. He classifies experiments on the basis of their role in inquiry as follows:[24]

(i) *E* is a 'crucial' experiment, i.e., decides between two or more theories or classes of theories. Franklin notes that such experiments are not as common as the textbooks of methodology might suggest, but mentions some examples of such experiments.

(ii) *E* is "strongly corroborative": *E* corroborates the basic or central ideas of a particular theory.

(iii) *E* exhibits a new phenomenon, unexpected on the basis of existing theories, and thus calls for the formulation of a new theory.

(iv) *E* tests an assumed anomaly which affects a certain theory *T*, and either confirms the existence of the anomaly, or shows that it is not genuine (but, for example, due to the malfunctioning of the experimental apparatus).

(v) *E* tests the experimental apparatus used in other experiments, and thus either confirms or disconfirms the results obtained in them.

Franklin does not discuss the reasons for performing various experiments, but rather the question what makes the experiments (cognitively) significant or important. Nevertheless he describes the experiments in terms of their function, at least in cases (i), (ii), (iv), and (v). The "crucial experiments" (i) are successful instances of question type (b1), whereas the "corroborative experiments" are presumably performed for the purpose of testing some theory (i.e., belong to (b2)). (Franklin notes that these experiments are sometimes difficult to distinguish from crucial experiments.) Experiments of type (v) are also instances of (b2), but in this case the relevant primary question is not a question about a theory, but about the trustworthiness of certain experimental results. Questions of this kind are addressed by the investigator to himself (or perhaps they are presented to him by other investigators); thus they must be regarded as primary questions. For this reason the primary questions cannot be identified with "theoretical questions" (in the ordinary philosophical sense of the word "theory"). The experiments in Franklin's category (iv) (those investigating anomalies) are also instances of (b1): they are performed for the purpose of resolving cognitive conflicts within belief system. Thus (b1) covers not only "crucial experiments" in the traditional sense of the word, but also other experimental questions whose purpose is to resolve conflicts between theories and observations (or experiments) or conflicts between different experiments.

VI

In the cases discussed above the purpose of an experimental question is to test an answer to some primary question (for example, a hypothesis or theory) or to identify the correct answer from a number of competing alternatives. This seems to be the "standard view" of the role of experiments in science among philosophers. Perhaps this assumption of the methodological primacy of theories with respect to experiments depends on the philosophers' preoccupation with the "context of justification" (as opposed to the "context of discovery"). However, it seems undeniable that even if experiments were performed for the purpose of testing a specific theory or hypothesis, they may suggest new hypotheses and lead to the discovery of new facts and theories, and thus help to find new answers to the investigator's primary questions. This is the case in Franklin's category (iii), in which an experiment reveals an unexpected phenomenon and calls for the formulation of a new explanatory hypothesis or theory: in the context of inquiry experimental questions of this kind can function as "discovery-questions", that is, as questions of type (a). There are also clear cases in which an experiment is performed for the purpose of finding (rather than testing) an answer to some primary question. A very simple case of this kind is one in which an investigator is interested in a theory involving some free parameters, and he performs experiments simply to determine the values of these parameters. Such experiments may simultaneously function as instruments of discovery and as means of testing a hypothesis.

The distinction between experimental questions of type (a) and questions of type (b) is related to Herbert A. Simon's distinction between "data-driven" and "theory-driven" science. In the latter, the data obtained by observation and experiment are used for testing a pre-existing theory (or theories), whereas a "data-driven" discovery process

> takes a body of data as its starting point and searches for a set of generalizations, or theory, to describe the data parsimoniously or to explain them.[25]

In his recent book *Representing and Intervening* Ian Hacking gives several examples from the history of optics in which observations (experimental questions) seem to have preceded any formulation of theory, and were made for the purpose of finding a theory for the phenomena under consideration (for example, Grimaldi's and Hooke's observations of illumination in the shadow of an opaque body, and Newton's observation of the dispersion of light).[26] Such examples abound in the history of psychology and other theoretically relatively undeveloped fields. For example, the recent experimental work on the psychology of decision-making provides good examples of experiments which serve the purpose of discovery. After it became evident that the Bayesian model of decision making under uncertainty is not a descriptively adequate theory of human decision making, a great many experiments have been performed for the purpose of identifying the factors which would account for the deviations from the Bayesian model. The principal aim of such experimentation has been the formulation of a good theoretical alternative to the Bayesian account of human decision making.

It is possible to make observations and even perform experiments solely out of curiosity, without reference to antecedent primary questions. (This does not prevent the observations from being guided by the investigator's theoretical beliefs.) In this case the experimental (secondary) question is also a primary question which the investigator poses to himself. But such questions are without theoretical significance (or "meaningless") as long as they remain unrelated to other primary questions.

University of Turku and University of Miami

NOTES

1. This paper is based on work supported by the U.S. National Science Foundation Grant No. IST-8310936 (Principal Investigators Jaakko Hintikka and C. B. J. Macmillan) and by the Finnish State Council for the Humanities Grant No. 09/053. Part of the work was done during the spring term 1986 while I enjoyed a research fellowship at the Center for Philosophy of Science, University of Pittsburgh. I am grateful to my Pittsburgh colleagues and especially to Jaakko Hintikka, Allan Franklin, Diderik Batens, and Nils-Eric Sahlin for discussions on the subject-matter of this paper.
2. R. G. Collingwood, *An Essay on Metaphysics*, Clarendon Press, Oxford 1940, p. 23.
3. Collingwood, *op. cit.*, p. 24.
4. This approach to the theory of inquiry has been explored in the articles published in *Synthese* 47 (1981), *Scientific Method as a Problem-solving and Question-answering Technique*; see especially Herbert A. Simon, Patrick W. Langley, and Gary L. Bradshaw, 'Scientific Discovery as Problem Solving', pp. 1–27; Jaakko Hintikka, 'On the Logic of an Interrogative Model of Scientific Inquiry', pp. 69–83; Thomas Nickles, 'What is a Problem That We May Solve It?', pp. 85–118; and Scott A. Kleiner, 'Problem Solving and Discovery in the Growth of Darwin's Theories of Evolution', pp. 119–192. See also Jaakko Hintikka, 'True and False Logics of Scientific Discovery', in *Logic of Discovery and Logic of Discourse*, ed. by J. Hintikka and F. Vandamme, Plenum Press, New York and London, 1985, pp. 3–14.
5. A good formulation of this view can be found in William James's essay 'The Will to Believe', in *The Will to Believe and Other Essays in Popular Philosophy*, Longman's Green and Co., London 1897, p. 17.

6. The expression "epistemic utility" was first used in this context by Carl G. Hempel; see his paper 'Inductive Inconsistencies', *Synthese* 12 (1960), pp. 439–469; reprinted in C. G. Hempel, *Aspects of Scientific Explanation*, Collier-Macmillan Ltd., London 1965.
7. For a discussion of truth and information as epistemic utilities, see Risto Hilpinen, 'Decision-theoretic Approaches to Rules of Acceptance' in *Contemporary Philosophy in Scandinavia*, ed. by R. Olson and A. Paul, pp. 147–168, and Isaac Levi, *The Enterprise of Knowledge*, MIT Press, Cambridge, Mass. and London, 1980, ch. 2.
8. This observation about beliefs was made by Frank P. Ramsey in 'Knowledge', in F. P. Ramsey, *The Foundations of Mathematics and Other Logical Essays*, ed. by R. B. Braithwaite, Routledge and Kegan Paul, London 1931, pp. 258-259. This "paradox" is analogous to the "Paradox of the Preface" discussed by David Makinson in 'The Paradox of the Preface', *Analysis* 25 (1964), pp. 205–207, and to the "Lottery Paradox"; for the latter, see Henry E. Kyburg, *Probability and the Logic of Rational Belief*, Wesleyan University Press, Middletown, Conn., 1961, p. 197.
9. For a discussion of the concepts of degree of falsity (or truth) and truthlikeness (verisimilitude), see Risto Hilpinen, 'Approximate Truth and Truthlikeness', in *Formal Methods in the Methodology of Empirical Sciences*, ed. by M. Przelecki, K. Szaniawski, and R. Wojcicki, D. Reidel, Dordrecht, 1976, pp. 19–42. Graham Oddie, *Likeness to Truth*, D. Reidel, Dordrecht 1986, contains a good general discussion of Popper's concept of verisimilitude and its contemporary varieties.
10. David Miller, 'On the Accuracy of Predictions', *Synthese* 30 (1975), pp. 159–191.
11. Miller's result has been interpreted in this way be Roger Rosenkrantz in his comment on Miller's paper, 'Truth-likeness: Comment on David Miller', *Synthese* 30 (1975), pp. 193–197.
12. *Collected Papers of Charles Sanders Peirce*, Vol. V, ed. by C. Hartshorne and P. Weiss, Harvard University Press, Cambridge, Mass. 1934, paragraph 5.584.
13. See Lennart Åqvist, *A New Approach to the Logical Theory of Interrogatives, Part I: Analysis*, Filosofiska Föreningen i Uppsala, Uppsala 1965, Introduction, section 2; Jaakko Hintikka, *The Semantics of Questions and the Questions of Semantics*, Acta Philosophica Fennica 28:4, North-Holland, Amsterdam 1976; 'Answers to Questions', in *Questions*, ed. by Henry Hiz, D. Reidel, Dordrecht 1975. pp. 279 - 300; 'New Foundations for a Theory of Questions and Answers', in *Questions and Answers*, ed. by F. Kiefer, D. Reidel, Dordrecht 1983, pp. 159 - 190.

14. See Jaakko Hintikka, 'Answers to Questions', in *Questions*, ed. by Henry Hiz, pp. 279-286.
15. I am referring here to the familiar "justified true belief analysis" of knowledge. Recent discussion has show that the conditions included in this analysis are not sufficient for knowledge, but they may nevertheless be regarded as necessary conditions. See Robert K. Shope, *The Analysis of Knowing: A Decade of Research*, Princeton University Press, Princeton, N.J. 1983, pp. 5-19.
16. Laurens Laudan, *Progress and Its Problems*, Routledge and Kegan Paul, London and Henley 1977, p. 15.
17. Laurens Laudan, *op. cit.*, p. 25.
18. Cf. Immanuel Kant, *Critique of Pure Reason*, transl. by Norman Kemp Smith, Macmillan, London 1964, p. 20 (B: xiii); Jaakko Hintikka, 'On the Logic of an Interrogative Model of Scientific Inquiry', p. 69; Lisa Jardine, *Francis Bacon: Discovery and the Art of Discourse*, Cambridge University Press, Cambridge 1974, pp. 136-137.
19. Jaakko Hintikka, 'True and False Logics of Scientific Discovery', in *Logic of Discovery and Logic of Discourse*, ed. by J. Hintikka and F. Vandamme, pp. 9–10, and 'What is the Logic of Experimental Inquiry?' (manuscript).
20. This objection to the epistemic semantics of questions was already raised by Bernard Bolzano, see his *Theory of Science*, ed. and transl. by R. George, University of California Press, Berkeley and Los Angeles, 1972, section 145 (pp. 192-193). According to Bolzano, a question is a request that a certain true proposition should be produced or "indicated"; Bolzano notes that such a request does not presuppose ignorance about the truth in question (on the questioner's part). (Op. cit., p. 193.)
21. Lennart Åqvist, 'Scattered Topics in Interrogative Logic', in *Philosophical Logic*, ed. by J. W. Davis et al., D. Reidel, Dordrecht 1969, pp. 114-121; see p. 120. Examination questions may be regarded as a subclass of *teacher's questions*; cf. Jaakko Hintikka, 'A Dialogical Model of Teaching', *Synthese* 51 (1982), pp. 39-59.
22. This concept of transmitted information has been discussed in Risto Hilpinen, 'On the Information Provided by Observations', in *Information and Inference*, ed. by Jaakko Hintikka and P. Suppes, D. Reidel, Dordrecht 1970, pp. 97-122 (cf. pp. 115-116).
23. Allan Franklin, 'Experiment, Theory Choice, and the Duhem-Quine Problem', in the present volume.

24. Allan Franklin, 'What Makes a 'Good' Experiment?', *The British Journal for the Philosophy of Science* 32 (1981), pp. 367–374.
25. See Herbert A. Simon et al., 'Scientific Discovery as Problem Solving', *Synthese* 47 (1981), p. 7.
26. Ian Hacking, *Representing and Intervening*, Cambridge University Press, Cambridge 1983, pp. 155–156. Cf. also Allan Franklin, 'The Epistemology of Experiment' (Review of I. Hacking, *Representing and Intervening*), *The British Journal for the Philosophy of Science* 35 (1984), pp. 381–401.
27. Nils-Eric Sahlin, 'The Significance of Empirical Evidence for Developments in the Foundations of Decision Theory' (in the present volume) contains a good discussion of this topic and additional references.

II

THE ROLES OF EXPERIMENT: THEORY GENERATION AND THEORY TESTING

THOMAS NICKLES

RECONSTRUCTING SCIENCE: DISCOVERY AND EXPERIMENT

And what the dead had no speech for, when living,
They can tell you, being dead: the communication
Of the dead is tongued with fire beyond the
language of the living.
- T. S. Eliot[1]

1. SINGLE-PASS VERSUS MULTI-PASS ACCOUNTS OF SCIENCE

Science transforms itself by more or less continuously reworking its previous results and techniques. To miss the dynamical, self-reconstructive nature of scientific work is to miss the extent to which scientific inquiry is a bootstrap affair. I shall call non-reconstructive views of science *single-pass* or *one-pass* models of scientific inquiry. Here are some examples of reconstruction.

In his book on the history of the early quantum theory, Kuhn (1978) argues that Planck did not advance a quantum theory in 1900. Rather, the quantum theory was invented by Einstein and Ehrenfest in 1905–7, who misinterpreted and reconstructed Planck's work, rooted in classical physics, as a radical solution (which Planck did not conceive and later repudiated) to a problem Planck did not have (the ultraviolet catastrophe).[2] By 1916 these and other physicists, such as Debye, Bohr, and Sommerfeld had provided new layers of constructive reinterpretation of "Planck's" quantum relations and more powerful derivations of them. By 1924 the spelling out of the Bose–Einstein statistics provided yet another deep reinterpretation of "Planck's" blackbody formula and theory. The later, canonical reformulations bore scant resemblance to the originally announced results.

Brannigan (1981) provides a remarkably similar story of the reconstruction of Mendel's work. According to Brannigan's socio-historical account, Mendelian genetics was a theory largely invented in 1900 by De Vries, Correns, and Tschermak, who cited and recharacterized Mendel's work partly as a means of avoiding priority disputes among themselves.

D. Batens and J. P. van Bendegem (eds.), Theory and Experiment, 33–53.

Third, consider the history of the Kepler problem (the two-body problem), from poor Kepler's struggles, to Newton's mathematical treatment in the *Principia*, and forward to how students are taught today in up-to-date mechanics textbooks. In fact, the history of most any standard textbook problem or of most any entry of a standard scientific handbook will illustrate my point. In successful cases scientists reconstruct ill-structured problems, noisy data, and weak, groping, solution methods as well-defined problem spaces to which strong search/solution methods can be more-or-less routinely applied. Contemporary examples may be found in Cushing (1982, 1986) and Pickering (1984).

The sort of conceptual and technical reconstruction I have in mind is *not* the rational reconstruction of philosophers, imposed on science from the outside. It is a process (actually many processes) largely *internal* to scientific practice, both theoretical and experimental. Reconstruction is not something that working scientists do merely for philosophical relaxation, while off duty. It is something that they must do, consciously or not, in order to apply old results and techniques to new problems at the frontier and to model one problem solution on another. Reconstruction is often necessary, for the foundations of major theoretical and experimental lines of inquiry are typically muddled at the beginning (Franklin, 1986; Galison, 1981, 1987; Hacking, 1983). Often scientists distill out the conceptual "essence" of previous work in terms that were not even available to the historical problem solvers. This process of applying/extending previous "knowledge" actually transforms that knowledge, sometimes beyond recognition. The process also "delocalizes" previous knowledge, to a greater or lesser extent, even if the result of the reconstruction is only to relocate it elsewhere.

The dominant philosophical accounts of science have been blind to the breadth and depth of scientific reconstruction. I shall say that they commit the *one-pass fallacy*. (To a lesser extent this is also true of historical and sociological accounts; see Nickles, 1987c.) Most philosophical models and methodologies of science have been one-pass conceptions, which force all scientific research at all times into the same mold. The traditional battle between simple inductivist and hypotheticalist methodologies is a case in point. Basically, these methodologies represent only two among many (families of) problem-solving strategies, but their many proponents have singled out one or the other of them and glorified it as a methodology adequate for all sciences at all times! (Popper is a recent example.) But why should all scientific investigation be inductive, or hypothetico-deductive? Indeed, why should experimental and theoretical research in the very same field be methodologically identical?

Failure to recognize the extent of conceptual transformation has even truncated explicitly constructivist accounts of science. For example, some constructivist sociologists too quickly agree with Sir Peter Medawar (1964) that the scientific paper is a fraud, on the ground that even experimental papers do not accurately report the previous research. But why should the paper be considered a mere historical report rather than an essential continuation of the research - an attempt to rework the problems and solutions in the logically strongest manner?

That working scientists themselves often have one-pass views of scientific work makes this issue all the more confusing. By not appreciating the distinction between "historical" mode of generation and "logical" *generatability* or *derivability*, scientists *do* sometimes write papers and textbooks disingenuously, as if their actual research proceeded in the logical order reported. A famous case is Newton's first published paper, on light and color (Newton, 1672; Nickles, 1985). The logic of Newton's presentation (a claimed deduction from the experimental phenomena) differs greatly from the logical character of the preceding work on which the paper was based. The authors of scientific papers recognize that while many papers (especially those concerning experimental work) are in some sense reports, it is legitimate to logically clean up the original investigation. However, it is rhetorically more effective (and self-flattering!) to present the research as if it were historically ordered in just the way its logical order suggests. Lakatos's research programs (which are partly retrospective reconstructions) do the same thing on a larger scale.

Although the two are continuous, I think of reconstruction as second-order construction and the overall project as a Phase II constructivist program: reconstructions rework previously developed material and techniques. Promising loci of study here are "historical" sections of papers and books (in which research programs are sometimes retrospectively outlined), the summary memoir, in which a researcher provides a "definitive," comprehensive summary of her previously reported results, and the technical review article, surveying the state of work in a problem area - not to mention *Handbücher* and textbooks. Most constructivists speak of construction out of resources which themselves remain unchanged (e.g., Edinburgh interest theorists and sometimes even Kuhn on exemplars), whereas reconstruction can transform those resources themselves. Consider how Planck's paper and theory were transformed by later consumers and appliers of it.

In ignoring the self-transforming character of scientific work, one-pass models can hardly avoid committing the *genetic fallacy* -

the mistake of thinking that its conditions of origin determine forever the character or "essence" of a thing. In fact, the *one-pass fallacy* is, in a sense, just the genetic fallacy or its opposite, which we might dub the *teletic fallacy* (from *telos*) - the fallacy of thinking that the origins of a thing must anticipate or "*pre*capitulate" its ultimate character. The beginning must presage the end. This is easy to see in the case of historical methodologies. If a certain experiment was not considered a crucial experiment by the scientific community which performed and licensed it, then it is not crucial *period*. If a problem solution originated as a conjecture supported by some predictive tests, then it remains forever a conjecture subject only to consequential support. If this is anti-whiggism, it is excessive.

The positivists were often guilty of the reverse, teletic fallacy. They allowed their interest in the logical structure of the products of research to dictate naive views about the nature of the research process itself. Since it is difficult to see how a logical structure can emerge gradually, or by parts, they tended to treat research activity, the *process* of research, as unstructured, both temporally and logically.

Both extreme formalists and extreme historicists have been guilty of the fallacy of *presentism* - the mistake of interpreting a historical period in terms of the present (sometimes a historical present!). But while historians, sociologists, and philosophers all have sliced time too thinly, it is especially the philosophers who have illicitly loved the instantaneous - the nearest historical counterpart to the timelessness of logic. Ironically, their blindness to the richly-textured domains of reasoning involved has eliminated opportunities for philosophical investigation. Many influential philosophers have treated the discovery of new scientific ideas as instantaneous and epistemically unstructured; experimental observation as instantaneous and unstructured; and even decision-making as nearly instantaneous - instant rationality. Things are permanent and fixed until, suddenly, they flip into another well defined state - like frogs on lilly pads or atomic state transitions. The dimension of time is not fully incorporated into such accounts, including Kuhn's.

One-pass views of science thus make the problem of understanding *historical change* intractable. Multi-pass treatments of science can provide alternative models of scientific revolutions as (usually gradual) conceptual reconstructions. Multi-pass models not only allow but predict revolutions in one perfectly good sense but without wrenching discontinuities.

2. THE METHODOLOGICAL SIGNIFICANCE OF RECONSTRUCTION

The reconstructive, multi-pass approach to science touches many central issues of methodology. Many of the topics thrown out by historical critiques of philosophical rational reconstructions come back in, only now as topics of working-scientific reconstruction rather than as something externally imposed by philosophers. Here I list several ramifications of reconstruction.[3] The rest of the paper leaves me space to expand on only a few of these topics.

1. The fact of reconstruction establishes the need for multi-pass models of science. Scientific research is inhomogeneous. There is no methodological unity of science. All research is *not* constituted by a single type of problem and, accordingly, by just one kind of solution strategy or solution structure. One and the same method need not be adequate to all problems met at all phases of research. A multi-pass approach, being nominalistic, avoids the danger of searching for methods relevant to all scientific problem solving. Only vapid abstractions can succeed in doing that.

2. Reconstruction bridges history, society, and, logic. Taking reconstruction seriously means providing more detailed and more genuinely historical and social-constructivist accounts of research. But it also means seeing how working scientists themselves construct logically clean representations of research results and their justification. Reconstruction simultaneously *divorces* logic from original history and *joins* them.

3. The divorce of original 'discovery' from justification (including reconstructed 'discoverability') is one example. The existence of reconstruction therefore reopens the question of logics of discovery and justification. Moreover, reconstruction is a major vehicle of scientific innovation and one that even the "friends of discovery" have overlooked (but see Finocchiaro, 1980). Understanding how it works can result in a less ad hoc account that conceives innovation as a normal part of the research process rather than as an exogenous factor, an extra-systemic mutation.

4. In particular, even if work at the frontier must often use weak methods, e.g., inductivist and hypotheticalist methods, those methods may later give way to more powerful problem-solving techniques, even mechanized ones. This shows the dangers of linking too closely original discovery or construction with "final" justification (Nickles, 1985). Generatability must not be confused with original generation.

5. Reconstruction implies that the logical status of scientific results can change significantly from one pass to another. Experimental data originally introduced within a consequential logic may eventually assume a role within a generative logic of justification. The "same" experimental results may be expressed as modest data claims, as somewhat stronger phenomenal claims or as bold, theoretical claims (cf. Pinch, 1984).

6. Reconstruction is relevant to the underdetermination problem, the problem that the data and other constraints are often insufficient to determine a problem solution. As the problem and its potential solutions are interactively transformed by reconstruction in the light of new information, a problem becomes better defined and its solution better determined and therefore better justified in relation to previous work. Of course, this is not a purely logical process. The logical framework is socially constructed.

7. One example is inconsistency claims, including falsification and the Duhem problem. Eventually the issues can become sufficiently settled that the community considers certain experimental information or the violation of a theoretical constraint to be fatal to a theory. Inconsistency claims have degrees of "externality" just as do observation claims (cf. Pinch, 1984). The fault may be assigned to the core theory itself rather than to an auxiliary assumption such as a parameter estimate or approximation.

8. Crucial experiments provide another example. On a multi-pass model, an experiment may be crucial on one pass (at one stage or level of construction) but not on another. Thus the fact that an experiment was not crucial for the people who performed it does not preclude its achieving crucial status later - and vice versa. E.g., in the early 1800's, many considered the existence of light radiation pressure a crucial test of particle versus wave theories of light, but by the time its existence was securely established a century later, the electromagnetic wave theory also predicted its existence, and the experiment had little confirmatory force (Worrall, 1982).

9. The status of experiments can change in other ways. A result which at one time is considered highly confirming may later be considered to carry little or no weight at all - or vice versa. An example is the role of gravitational red shift in the history of relativity theory (Earman and Glymour, 1980). The set of predictions attributed to a theory or type of theory may change non-

cumulatively over time as the logical linkages and auxiliary assumptions are sorted out differently.

10. Where controversy once existed, experimental closure can be viewed as an instance of scientific reconstruction. Experimental closure resembles theoretical closure (esp. generative justification of theory) in important respects, and replication amounts to an experimental variety of discoverability. Conversely, discoverability is a form of theoretical replicability.

11. A reconstructive view of science permits us to resurrect the idea of "positive science" - a body of reliably established results, over against the familiar, 20th-century, fallibilistic viewpoint that since all scientific claims are conjectural or hypothetical, we must totally abandon the old view that what begins life as a conjectural hypothesis may ultimately achieve a different logico-epistemological status. Such ideas, suitably qualified, are necessary to make sense of almost every aspect of scientific research.

12. There is no point in seeking a single type of scientific "rationality." While some decisions must be made at the moment, on the basis of limited and unreliable information, other judgments are, in effect, transgenerational (Will, 1981). Just as philosophers have collapsed history into a time-slice and scientific discovery into a momentary "aha" experience of individuals, so they have telescoped rational decision-making into the "instant rationality" of individuals.

13. Reconstruction bears crucially on issues of scientific change and sources of innovation. E.g., the issue is *not* simply whether or not scientific change is cumulative, where the units of information or achievement are conceived as little bricks which are added or removed. For these units are themselves transformed in the assimilation process.

3. GENERATIVE VS. CONSEQUENTIAL LOGICS OF EXPERIMENT

A particular application of the reconstruction point is the distinction between discovery and discoverability which I have introduced elsewhere (Nickles, 1985, 1987a and b). This point involves the distinction of generative from consequential logics of justification. The general moral here is that once we attend to the context of discovery and to the way in which the first, primitive problem solutions are conceptually reconstructed over time, even our view of "final" justification is transformed into something very different from positivistic and Popperian accounts. The point of adequately treating "discovery" is not simply to fill a lacuna in received accounts of science.

Classical *generativists* like Bacon and Newton held that the epistemically superior theories are those constructed or derived from what we already know about the world - the phenomena plus other previously established results. The basic, 17th-century, methodological conflict set generativists against *hypotheticalists* or *consequentialists*, who held that empirical support results from successfully testing the predictive consequences of theories, no matter how those theories were obtained. Today most philosophers of science believe that the consequentialists won the battle: only the consequential testing of theories provides empirical support.[4] The *central dogma* of hypothetico-deductive (H-D) methodology is that

> All empirical support = empirical evidence = empirical data = successful test results = successful predictions = true empirical consequences of the claim (plus auxiliary assumptions).[5]

Popperian and Lakatosian methodologists constitute an extreme but important example of modern-day consequentialism. On their view only "novel" predictions provide support (or corroboration). Hence, no information used to construct a theory can possibly count in its support (Worrall, 1978).

This *novel* consequentialist view of experimental justification is precisely the opposite of the extreme (anti-novelty) generativist view, according to which *only* the evidence used to construct the theory counts in its support. I do not know of any historical figures who were this extreme, but the stereotype of classical inductivism amounts to this position: the only admissible scientific theories are those which are derived from the known facts. Strong inductivists attach relatively little importance to predictive consequences, whereas novel consequentialists deny that inductive information provides any support at all. Because of this logical opposition, I term the change from predominantly generativist to consequentialist methodologies *the great logical inversion*. This inversion is supposed to have occurred in the 19th century.

On my reading (Nickles, 1987d), the history of science (e.g., the history of physics) does not reveal an abandonment of generative methodology by working scientists. The change is most evident in philosophical writings about science, but even here it has been exaggerated.

Since discoverability arguments *are* arguments, generative justification is a *logical* rather than a *historical* theory of justification (Musgrave, 1974) and thus avoids the problems with historical theories (What did Einstein know and what was in his mind when he...?). But insofar as our theory of justification is logical rather than historical, and insofar as science typically is a reconstructive, multi-pass enterprise, it follows that the logical, justificatory status of claims may change historically. This includes experimental claims.

By now it should be apparent that I see no philosophical justification for abandoning generative justification. In recent papers I have defended generativism as a form of justification complementary to consequential testing and sometimes logically stronger than the latter.[6] Generative justification furnishes sufficient conditions for claims rather than the necessary conditions that H-D justification provides. A theory can be justified directly by reasoning *to* the theory from what we already 'know' as well as indirectly by reasoning *from* the theory plus auxiliary assumptions to predictive consequences. Generative justification is constructive; consequential testing is purely eliminative.

4. VARIATIONS ON THE GENERATIVE LOGIC OF EXPERIMENT

Generative argument cannot derive universal claims from particular claims or high-theoretical claims from lower-level phenomenal premises, of course[7]; however, generative justification can be partial and local. Among the premises can be a variety of previously established results. Indeed, the derivation of a claim may involve no phenomenal premises at all. But since our topic is experiment and theory, I shall here ignore purely theoretical derivations.

Within a given context of reliable results, phenomenal claims can have important consequences. For example, from a mathematical generalization of the Galilean transformation together with the constancy of the velocity of light (here considered a phenomenal claim), one may derive the Lorentz transformation.[8] Second, Ehrenfest (1911) derived the existence of energy quanta from Planck's *empirical*, black-body radiation law in conjunction with some unproblematic generalizations of classical statistical thermodynamics.[9] By 'generalizations' in both cases I mean not logically stronger claims but logically weaker abstractions of the original principles. Most instances of the generative use of experimental information are not so dramatic, of course. For instance, data may be employed to fix a free paramater of a model or theory.[10]

In the history of science we frequently find the following sort of situation. Experimental information which originally has a consequential use eventually acquires a generative role.[11] This happens whenever the experimental information is employed not merely to test an independently-proposed theory candidate but, more actively, to impose a formal condition on any adequate theory. This is what Ehrenfest did in great detail in his study of Planck's empirical law.

The logical status of experimental work can change in other ways. Earman and Glymour (1980) report that the existence of a gravitational red shift (of the right magnitude) was an important experimental question during the heuristic-developmental stages of general relativity. Red shift also provided an important test during the early stages of the completed theory's reception, when the experimental observations appeared (to some) to refute the theory. However, once the theory was better understood, it was discovered that red shift imposed no constraints on the theory not already imposed by the condition that Newton's equations remain valid to the first order. (Note the switch to a generative mode of thinking here.) Hence, red shift is no longer considered an important confirmation of the theory.

It is obvious by now that the logical role of experimental results in scientific inference is highly context-sensitive. It is especially sensitive to the background of previously established results, including theoretical results. Philosophers and philosophical historians have sometimes written as if all scientific results are underdetermined by logic and the data (or the phenomenal claims which the data support). In a global sense, this is perhaps true; but within the context of a given theoretical-experimental background, alternative choices may reduce to a small number. In such a case further experimental work may suffice to eliminate all but one of these candidates. In such local research contexts, experiment can assume the role of classical 'eliminative induction' (which - apart from computational fudges - is really a form of deduction, given the background). This amounts to using experimental information in a sort of *indirect* generatability argument, given that the context cuts down the huge space of logically possible hypotheses to the subspace containing the few remaining, plausible alternatives. In certain cases the mechanization of experimental problem-solving and theoretical computation has proven possible.

A weaker sort of generative use of empirical evidence is illustrated by Pinch's (1981, 1985) discussion of the solar neutrino problem. The trouble here is that the observed level of solar neutrinos is far below the predicted range of values. However, the calculations are exceedingly complex, involving many approximations, simplifications, parameter estimations, and assumptions of reliability in several distinct areas of physics, astronomy, and experimental technology. The disappointingly low results led Bahcall and other scientists to search mainly for ways in which to adjust the parameters so that a lower theoretical prediction would be obtained. The experimental results constituted *prima facie* evidence, given the context, that the "lower" values of the parameters were more likely to be correct. The hope was thereby to avoid a contradiction between theory and experience. The heuristic search strategy was to assume as a premise that no genuine (highly external) contradiction exists and then let the theory and phenomena (via complex series of calculations) select those parameters that would require adjustment. Furthermore, insofar as the adjustments fell within otherwise acceptable limits, this procedure would, to some extent, justify those adjustments. What we really have here in this mutual adjustment of theory to phenomena is *parameter fixing* of a sophisticated variety. Such *mutual* adjustment goes beyond the bounds of strictly consequential thinking.

Clearly, experimental information plays a strong generative role in much experimental reasoning itself. This may seem too obvious to mention, but let us recall that on recently standard philosophical views about observation and experiment there was no possibility of interesting reasoning in experimental contexts at all. One simply manipulated apparatus, observed, recorded one's protocols and tried to get them accepted by the scientific community as new "basic statements." This view was a ridiculous oversimplification of experiment during any epoch. As Galison (1987) points out, in modern experimentation the physicist herself does little actual manipulating and looking. Virtually all of her work is concerned with design questions and with modes of data reduction and analysis. In fact, experimental contexts contain the resources for reasoning of complex and subtle kinds. In this respect, there is a parallel between positivist-Popperian treatments of experiment and discovery. In both cases the contexts were supposed to be impoverished of the resources necessary to construct reasoning of philosophically interesting varieties.

So far my talk of the generative use of experiment may suggest that experimental data are commonly used as premises in theoretical derivations. Actually, matters are more complicated. In fact, I have nearly always spoken of "phenomenal" premises rather than "data" premises. Following Bogen and Woodward (forthcoming) let us distinguish *data*, the "local" information produced by particular experiments from *phenomena*, the empirical "effects" or regularities more globally claimed to exist in nature. Now, roughly speaking, the reasoning linking theory and experiment divides into two levels. (In many cases there is a whole hierarchy of levels, as Bogen and Woodward point out.) At one level theories are related to phenomenal claims in explanatory, predictive, and also generative ways. We explain phenomena, not data. At another level, phenomenal claims are related to the experimental data. I agree with Bogen and Woodward that since the data depend heavily on a host of local contingencies reflecting the experimental design, the nature of the apparatus, the particular conditions which obtained, and whatnot, the data do not constitute a homogeneous domain which it is the theory's job to explain. It is usually pointless to seek a general theory of the data. Rather than providing explananda or "predictanda" for theory, the data are employed epistemically to establish the existence and nature of phenomena. As before, the reasoning can proceed in any direction, but it would be incredible to deny that generative reasoning has a major role here.[12]

My conclusion is that not all scientific reasoning is governed by simple, hypotheticalist logic. Global hypotheticalism *cum* underdetermination do not entail hypotheticalism in local research contexts. This global-local fallacy combines with the one-pass fallacy to work much mischief in science studies.

5. CRUCIAL EXPERIMENTS

I end by applying the reconstructionist and generative perspectives to crucial experiments. Here we find a by now familiar pattern of thinking.

Historians have discovered that some of those experiments commonly termed crucial had no special significance for the historical agents who originally performed or received them. Examples are Young's interference and diffraction experiments (if he performed them at all), the Fresnel-Arago spot experiment, the Foucault speed-of-light experiment, and the gravitational red-shift.[13] But so what? An experiment originally considered to be crucial may later be denied this status by scientists themselves, on the basis of their reconstruction of this domain of knowledge. Similarly, reconstruction may bring about a shift in the logic of experiment such that a once-ordinary experiment, in the reconstructed context, becomes crucial in deciding among leading candidates for a problem solution. Failing to admit this possibility is an error almost as serious as holding that an empirical anomaly facing a new theory will forever remain anomalous, that all anomalies are refutations. It confuses historical reports with logical claims or at least fails to allow for historical change in perceived logical status. The original history of an experiment does not preclude a later conceptual reconstruction on the part of scientists. The reconstructions are themselves made in historical time, of course, and are subject to revision.

So here the first of three distinctions becomes necessary, a distinction between *logically crucial experiments* and *historically-socially-psychologically crucial experiments*. In the latter sense an experiment is crucial if it is decisive in converting the scientific community to one theory over its competitor(s). In the logical sense an experiment is crucial if (under a given reconstruction of the experiment, the domain of knowledge, and the competing theories) it entails the epistemic superiority of one theory over a competitor. The two senses are not exclusive, of course. However, it is difficult to find pure, logical cases in the original, historical "pass," because of Duhem's point that there are always several auxiliary assumptions that can be adjusted to logically save a given theory and Lakatos's point that the theories involved are often merely stepping stones within a larger research program.[14]

To illustrate, consider Fizeau's 1852 experiment on the velocity of light in running water. As Hacking (1983, p. 239) observes, Fizeau himself could not possibly have intended this experiment as a crucial test of the velocity addition "law" of special relativity versus the classical addition law. Yet Einstein in 1916 spoke of "the conclusiveness of the experiment as a crucial test in favour of the theory of relativity."[15] Here we have a putative case in which reconstruction plus significant theory change made an experiment crucial between viewpoints one of which did not and could not have existed when the experiment was first performed. Of course, scientists undertake such reconstruction/rehabilitation only when it is relevant to a contemporary problem. Working scientists seem less interested than philosophers in studying in foundational questions for their own sake.

As with our other topics, recent discussion has considered crucial experiments only as a variety of consequential testing. A crucial experiment confirms one of two competing theories and falsifies the other. In the context of the auxiliary assumptions employed, its results are necessary for the truth of the successful theory and sufficient for the falsity of the other.

Here a second distinction of two types of crucial experiment needs to be made. It follows from our previous discussion that there can be *generative crucial experiments* as well as the *con-*

sequential crucial experiments. Further, a *directly generative* crucial experiment provides information which, in the context of the other constraints, is *sufficient* to generate one of the competing theories (sufficient for its truth and thus also sufficient, in that context, for the falsity of a competitor). Recall Ehrenfest's use of Planck's empirical law to establish the reality of energy quanta.[16] This is a case in which experimental information is strong enough to impose a formal condition on theories, a condition which all theories but one fail and which helps generate the "correct" theory.

An *indirectly generative* crucial experiment falls between a purely generative and a purely consequential crucial experiment. Given the enriched background, which amounts to a constrained generator of all viable problem solutions (or at least an eliminator of virtually all occupants of the solution space), the experimental information is sufficient to refute all competitors but one. So far it sounds like an ordinary, consequential crucial experiment. However, relative to the context, it is sufficient for the truth of the surviving theory as well as sufficient for the falsity of the others. "Airtight cases" such as the Vine-Matthews-Morley confirmation of plate tectonics approach this happy state.

Bogen and Woodward's distinction of phenomenal from data claims motivates my third and last major distinction. On the one hand, we may speak of a *crucial phenomenon*, the existence (or nonexistence) of which is decisive between two theories or programs. On the other hand, we may speak of *crucial data* - experimental results which, in context, are sufficient to decide among phenomenal claims. In this sense crucial data demonstrate the (non) existence of a debated phenomenon, regardless of the import of this phenomenon in deciding among theory claims. Let us call this sort of experiment a *demonstrative experiment*. This is a "purely experimental" or "nontheoretical" concept of crucial experiment. *It is not only high theoretical debates which make experiments crucial*.[17]

A *crucial experiment* full stop is (roughly) a two-stage affair. It is an experiment which demonstrates the (non)existence of a crucial phenomenon *and thereby* decides among contesting theories. (Actually, there may be a hierarchy of stages, corresponding to several degrees of externality.) Both stages may be exceedingly difficult to establish, and *either one may occur in the absence of the other*. When light pressure was a crucial phenomenon, no one could demonstrate it. By the time a demonstration became possible, it was no longer crucial (at least in a qualitative sense). Interference and diffraction were known to Grimaldi and Newton, but at the time they were not crucial phenomena between the classical particle and wave theories of light (since at least the latter did not yet exist).

In my terminology, the work of Paschen and other experimentalists before the turn of the 20th century was crucial in demonstrating the *phenomenon* of the exponential drop-off of black radiation at high frequencies (the Wien limit). This phenomenon, expressed by the exponential form of the conjectural Wien-Paschen distribution "law" and later by that of Planck's empirical law, crucially established the existence of energy quantization, as Ehrenfest (1911) later showed. Thus Ehrenfest's reconstruction (via Planck's empirical law, which also encompassed the low-frequency limit results of Lummer and Pringsheim, Rubens and Kurlbaum that eventually led to wave-particle duality) in effect showed that the high frequency measurements were crucial in establishing the existence of energy quantization. At the time of their work, of course, none of these experimentalists had dreamed of energy quantization.

Historical and sociological critics have shown that standard methodological terminology is misleadingly punctiform - treating results as if they were achieved instantly. Since I earlier criticized lovers of the instantaneous, I must concede that "crucial experiment" like "discovery" usually has been treated as event-like. This is not surprising, for a crucial experiment *is* an experimental 'discovery'. Just as historians have shown that discoveries can only rarely be historically pinpointed as to place, time, and agent (Kuhn, 1962; Schaffer, 1986), so then for crucial experiments. Novel experimental work rarely wears its meaning on its sleeve (Gooding, 1986). Definitive work usually builds on a succession of previous efforts. The black-body case well illustrates this historical point. Depending on how much quantitative precision and interpretative sophistication is desired, the work of Paschen, Lummer and Pringsheim, and Rubens and Kurlbaum was anticipated by earlier workers, notably Langley but others, too - H. F. Weber, Balfour, and even W. Herschel and J. H. Muller (Kangro, 1972). Some of these investigators worked in an astrophysical (especially solar radiation) tradition; Rubens may be considered Langley's successor in important respects; and so on. It would be silly to pick out just one experiment by just one of these men (or teams) and declare *that* to be the crucial experiment which (as later reconstructed) establishes the existence of energy quanta. But even when we cannot pick out a crucial experiment in the historical-psychological sense, it may still make sense to pick out a *phenomenon* as crucial (relative to a reconstruction).

Let us remember, then, that experiments, like theories, undergo refinement and reconstruction. The word 'experiment' is multiply ambiguous between a particular historical event; an entire series of localized events; a type of procedure that may be performed repeatedly, by many persons in many locations by a (limited) variety of means; and full-fledged experimental traditions. When it is claimed in a logical sense that a certain experiment is crucial, what is properly meant is that a refined and reconstructed experiment of the type attributed to historical agents, provides, for the relevant community, a clear demonstration of the crucial phenomenon. Einstein (1916, p. 41) speaks of

> a most important experiment which the brilliant physicist Fizeau performed more than half a century ago, and which has been repeated since then by some of the best experimental physicists, so that there can be no doubt about its result.

Department of Philosophy
University of Nevada - Reno

NOTES

Acknowledgment. This paper draws on work supported by the U. S. National Science Foundation. I am also indebted to the John Dewey Foundation for research support.

1. Eliot (1943, p. 32). Quoted by Polanyi (1962), p. 69. I adopt the poet's words to my purpose. The communication Eliot had in mind was prayer and reflection on the historical meaning of a place. Nonetheless, there is a parallel. Eliot dwells on the intersection of time and eternity. My more prosaic topic is the intersection of history and logic.
2. See Kuhn (1978). On the ultraviolet catastrophe, see Klein (1962, 1970).
3. For additional ramifications of reconstruction, especially for sociology of science, see Nickles (1987a and b.)
4. Bayesian confirmation theorists, with their emphasis on the importance of prior probabilities, represent a growing body of opposition to the strict consequentialist view. I am not a Bayesian in a broad, methodological sense, but I agree that their analyses are quite useful in local, well-defined contexts.

5. So stated, the central dogma harbors the common, fundamental confusion of phenomena and experimental data (Bogen and Woodward, forthcoming). See below.
6. Not always stronger, because generative justification is often partial. My defense of generative methodology is not by any means a return to classical inductivism. Inductivism is just one species of generative support, and I am far more interested in other types. When a theory is derivable from previously established results, for example, the support is generative, even though there is no inductive step in the argument. Generativism ≠ inductivism. For many examples and a suggestive logical schema, see Dorling (1973).
7. However, many phenomenal claims are universal and some are pitched at a highly theoretical level. Thus the standard terminology (which I employ here) misleadingly suggests that 'phenomenal' and 'theoretical' are exclusive terms. Also, derivations need not be strict deductions. Because of the numerous approximations, simplifications, and reliability judgments necessary, modern theoretical and experimental reasoning usually is not strictly deductive. By inclusion of strong heuristics, problem-solving methods can be highly routinized even though not reduced to ironclad, deductive rules.
8. See Spector (1972, Ch. 10) for a clear, accessible discussion.
9. Compare Planck's own "backwards" derivation of his crucial entropy formulas in 1899-1900, starting from the empirical laws (Wien's law at first, then his own) as premises rather than conclusions of the reasoning.
10. See Nickles (1987a, §5) for a critique of the Lakatosians on parameter fixing.
11. For examples of this pattern, see the discussions of Newton's 1672 optical paper and Herschel's methodology in Nickles (1985, 1987d).
12. As Bogen and Woodward point out, there is an asymmetry between experiments which do and those which do not establish the existence (or nonexistence) of phenomena. A successful experiment establishes the (non)existence of a phenomenon which theories of the relevant domain must handle, but an experiment which fails to clearly establish the existence or nonexistence of a phenomenon (or feature of a phenomenon) places no demands on theory. A theory is not required to explain the data of defective experiments. Indeed, the theory itself does not explain the data in either case. I believe that this asymmetry provides one mechanism for retaining theoretical closure in science.

13. On Young, Fresnel-Arago, and Foucault, see Worrall (1976, 1987). More relevant to my logical point are historical claims that the work could not have constituted a logically crucial experiment, given what we know now. Worrall makes some comments of this sort on Newton's optical work (Worrall, 1983) as do Farley and Geison (1974) on the Pasteur-Pouchet controversy and Earman and Glymour (1980) on the red shift. Of course, crucial experimental claims, like all others, are fallible.
14. See Lakatos (1974). Lakatos repeatedly denied that genuinely crucial experiments occur in science. What occurs are wars of attrition between research programs. What we call crucial experiments, he said (correctly for most cases), are retrospective reconstructions rather than punctiform events; but he conceived crucial experiments as philosophical reconstructions of no scientific interest. Ironically, Lakatos himself made research programs too punctiform in origin, since all the 'passes' are laid out in advance, linearly, with no dead ends. Lakatos's research programs are retrospective reconstructions - collapsed histories.
15. Not only in a popular book, Einstein (1916, p. 41) but also in more technical contexts. Michael Redhead saved me from a slip here.
16. We may speak here of crucial data rather than a crucial experiment. Or we may speak of the experiments which confirmed the law as crucial.
17. Naturally, lower-level theoretical assumptions of many sorts are typically involved in all experimentation. My point here, in agreement with Hacking (1983), is that experimental work has its own goals, problems, and standards and does not exist only to decide theoretical questions.

REFERENCES

Bogen, J., and Woodward, J. (forthcoming). Saving the Phenomena. *Philosophical Review*.

Brannigan, A. (1981). *The Social Basis of Scientific Discoveries*. Cambridge: Cambridge Univ. Press.

Cushing, J. (1982). Models and Methodologies in Current Theoretical High-Energy Physics. *Synthese* 50, 5-101.

Cushing, J. (1986). The Importance of Heisenberg's S-Matrix Program for the Theoretical High-Energy Physics of the 1950's. *Centaurus*. In press.

Dorling, J. (1973). Demonstrative Induction. *Philosophy of Science*, 40, 360-372.

Earman, J. & Glymour C. (1980). The Gravitational Red Shift as a Test of General Relativity: History and Analysis. *Studies in History and Philosophy of Science*, 11, 175-214.

Ehrenfest, P. (1911). Welche Züge der Lichtquantenhypothese spielen in der Theorie der Wärmestrahlung eine wesentliche Rolle? *Annalen der Physik*, 36, 91-118. Reprinted in *Paul Ehrenfest: Collected Scientific Papers* , ed., M.J. Klein, pp. 185-212. Amsterdam: North-Holland Pub. Co., 1959.

Einstein, A. (1916). *Relativity: The Special and General Theory*. 15th edition, 1952. New York: Crown Publishers.

Eliot, T.S. (1943). *Little Gidding*, the last of his *Four Quartets*. New York: Harcourt, Brace.

Farley, J. & Geison, G. (1974). Science, Politics and Spontaneous Generation in Nineteenth-Century France: The Pasteur-Pouchet Debate. *Bulletin of the History of Medicine* , 48, 161-198.

Finocchiaro, M. (1980), Scientific Discoveries as Growth of Understanding: The Case of Newton's Gravitation. In Nickles, ed. (1980), pp. 235-255.

Franklin, A. (1986). *The Neglect of Experiment*. Cambridge: Cambridge Univ. Press.

Galison, P. (1981). Kuhn and the Quantum Controversy (review of *Black-Body Theory*). *British Journal for the Philosophy of Science*, 32, 71-85.

Galison, P. (1987). Four Histories of Experiment. In Gooding *et al.* (1987), in press.

Gooding, D., Pinch, T., and Schaffer, S., eds. (1987). *The Uses of Experiment*. Cambridge: Cambridge Univ. Press.

Gooding, D. (1986). How Do Scientists Reach Agreement about Novel Observations? *Studies in History and Philosophy of Science*, 17, 205-230.

Hacking, I. (1983). *Representing and Intervening*. Cambridge: Cambridge Univ. Press.

Kangro, H. (1976). *Early History of Planck's Radiation Law*. New York: Crane, Russak.

Klein, M. (1962). Max Planck and the Beginnings of the Quantum Theory. *Archive for History of Exact Sciences*, 1, 459-479.

Klein, M. (1970). *Paul Ehrenfest: Theoretical Physicist*, Vol. 1. Amsterdam: North Holland.

Kuhn, T.S. (1962). *The Structure of Scientific Revolutions*. 2nd ed., enlarged, 1970. Chicago: Univ. of Chicago Press.

Kuhn, T.S. (1978). *Black-Body Theory and the Quantum Discontinuity*. Oxford: Oxford Univ. Press.

Lakatos, I. (1974). The Role of Crucial Experiments in Science. *Studies in History and Philosophy of Science*, 4, 309-325.

Medawar, P. (1964). Is the Scientific Paper Fraudulent? *Saturday Review* (August 1), pp. 43–44.
Musgrave, A. (1974). Logical versus Historical Theories of Confirmation. *British Journal for the Philosophy of Science*, 25, 1–23.
Newton, I. (1672). A Letter of Mr. Isaac Newton ... Containing his New Theory about Light and Colors. *Philosophical Transactions*, 6, 3075-3087.
Nickles, T., ed. (1980). *Scientific Discovery, Logic, and Rationality*. Dordrecht: Reidel.
Nickles, T. (1985). Beyond Divorce: Current Status of the Discovery Debate. *Philosophy of Science*, 52, 177–207.
Nickles, T. (1987a). Lakatosian Heuristics and Epistemic Support. *British Journal for the Philosophy of Science*, 38, in press.
Nickles, T. (1987b). Justification and Experiment. In Gooding *et al*. (1987), in press.
Nickles, T. (1987c). The Reconstruction of Scientific Knowledge. Forthcoming in *Philosophy and Social Action*.
Nickles, T. (1987d). From Natural Philosophy to Metaphilosophy of Science. In *Theoretical Physics in the 100 Years since Kelvin's Baltimore Lectures*, ed. P. Achinstein & R. Kargon, pp. 507–541. Cambridge, MA: MIT Press.
Pickering, A. (1984). *Constructing Quarks*. Chicago: Univ. of Chicago Press.
Pinch, T. (1981). The Sun-Set: The Presentation of Certainty in Scientific Life. *Social Studies of Science*, 11, 131–158.
Pinch, T. (1984). Towards an Analysis of Scientific Observation. *Social Studies of Science*, 15, 3–35.
Pinch, T. (1985). Theory Testing in Science. *Philosophy of the Social Sciences*, 15, 167–187.
Polanyi, M. (1962). The Republic of Science. *Minerva*, 1, 54–73.
Schaffer, S. (1986). Scientific Discoveries and the End of Natural Philosophy. *Social Studies of Science*, 16, 387–420.
Spector, M. (1972). *Methodological Foundations of Relativistic Mechanics*. Notre Dame, IN: Univ. of Notre Dame Press.
Will, F.L. (1981). The Rational Governance of Practice. *American Philosophical Quarterly*, 18, 191–201.
Worrall, J. (1976). Thomas Young and the 'Refutation' of Newtonian Optics ...., In *Method and Appraisal in the Physical Sciences*, C. Howson, ed. Cambridge: Cambridge Univ. Press, pp. 107–179.
Worrall, J. (1978). The Ways in which the Methodology of Scientific Research Programmes Improves on Popper's Methodology. *In Progress and Rationality in Science*, ed. G. Radnitzky and G. Andersson, pp. 45–70. Dordrecht: Reidel.

Worrall, J. (1982). The Pressure of Light: The Strange Case of the Vacillating Crucial Experiment. *Studies in History and Philosophy of Science*, 13, 133-172.
Worrall, J. (1983). Hypotheses and Mr. Newton. Lecture, to be published.
Worrall, J. (1987). The Role of Successful Predictions in Theory Change. In Gooding *et al.* (1987), in press.

ERWIN HIEBERT

THE ROLE OF EXPERIMENT AND THEORY IN THE DEVELOPMENT OF NUCLEAR PHYSICS IN THE EARLY 1930'S

I. INTRODUCTION

The immediate context of the discovery of nuclear fission in late 1938 is connected with the experimental investigations of Otto Hahn and Fritz Strassmann at the Kaiser Wilhelm Institut in Berlin. The theoretical interpretation and experimental verification of fission was provided jointly at that time by Otto Robert Frisch and Lise Meitner while they were living in Copenhagen and Stockholm respectively. The discovery was totally unexpected and spectacular, but accepted immediately by scientists everywhere.

In view of the enormous experimental and theoretical advances accomplished in nuclear physics prior to late 1938, it has seemed worthwhile to examine in some detail the historical circumstances that might have served to set the stage for the discovery of fission. Unfortunately, there seems to be virtually no evidence, barring hindsight, to support the view that nuclear theory, in 1938, was prepared to accommodate anything akin to nuclear fission. Nor do we find, even after the discovery, that a convincing theoretical rationale for the phenomenon was forthcoming except in an *ad hoc* and makeshift way. At most, it is conceivable to maintain that nuclear theory served, in some subterranean or poking way, to influence the direction that experimentation took, which then brought to light the phenomenon of nuclear fission.

An examination of the reciprocity that maintains between experiment and theory in the pre-history, and discovery, of nuclear fission, provides the historian of science with a case study singularly skewed toward the dominant significance of experiment over theory. The objective in this inquiry is to show that the circumstances associated with events and discoveries leading to nuclear fission were predominantly related to questions of experimental strategy, instrumental feasibility, and the styles and modes of research practiced in various national laboratories. At most, adjacent theoretical endeavors provided no more than rhetorical stance or epistemological claim. At that time, they mostly seemed to be irrelevant or at least peripheral to what was going on at the frontier of the discipline. Sometimes they simply evaporated

D. Batens and J. P. van Bendegem (eds.), Theory and Experiment, 55–76.

as experimental advances were made. The essential point, however, is not to overlook or obscure the fact that the theoretical advances that were being made were of seminal importance for the discipline; and this in spite of the historical evidence that the correlation of experiment with theory in the 1930's seemingly got nowhere. The nucleus, one might say, turned out to be a harder nut to crack than the experiments or the theory showed, and certainly more complex than any books or papers written on the subject.

We shall see that the discovery of new information concerning the nucleus, was to fall heavily upon laboratory training and expertise, experimental finesse, and a knack for the invention and ability to design and re-design instruments that would provide answers to specifically focused empirical questions.

The experimentalists, mostly, lived in an intellectual and geographical space of their own making. It was a world dominated by their own inventions, and one which, so-to-speak, revealed from within itself the possibilities and challenges that were manageable, or almost so. That is, hurdles were manageable, provided enough ingenuity and craftsmanship were at hand to conceive, build, and put together the necessary equipment. It was a microcosm in which what had been accomplished served as the driving force for what might be accomplished. Puzzlements and resolutions spontaneously just led to new puzzlements. The Austrian dramatist Franz Grillparzer has provided the *bon mot* for our experimentalist: "where riddles led me on to further riddles; to them the truth was quite precisely known."

II. NUCLEAR PHYSICS BEFORE 1930

An historical investigation of events in nuclear physics that would cover the decade of the 1930's, in a creditable way, is too ambitious an undertaking on this occasion. Accordingly, only the most pertinent and most actively studied experimental and theoretical problems have been touched upon that are covered by the literature until about the mid-1930's. It was at that time that neutron-irradiation experiments with the heaviest elements - that eventually led to the discovery of nuclear fission - were hotly pursued, but not resolved, in various laboratories around the world. An analysis of the details, intricacies, and contextual circumstances that culminated in the discovery, itself, necessarily will have to be treated on another occasion.

The pre-history of this branch of nuclear physics and chemistry is an inordinately complex and untidy subject. Even after the discovery of fission had been made, verified, and universally accepted, explanations for the phenomenon in terms of the details, specific circumstances, and proposed theoretical models for the reaction, were in an unsettled state for decades. By contrast, the military, and subsequently, the civilian exploitation of the discovery were promoted at a relatively rapid pace. This not only distorted public perception of the role of the physicist in modern societies, but served to launch nuclear weapons and nuclear power into the arena of international politics. That being the case, it is pertinent to indicate that nuclear fission is but a subdomain within nuclear physics, which itself in turn is a rather small subdiscipline within physics.

Mention of a few landmarks prior to the discovery of the neutron in 1932, will serve to set the stage for a closer examination of the golden age of nuclear physics in the 1930's. At the 1978 University of Minnesota Symposium on nuclear physics in the 1930's, Hans Bethe, in his talk on "The Happy Thirties," commented that his title was "perhaps a little bit unusual because politically the thirties were anything but happy." Decades before the "thirties," major experimental advances in nuclear physics and chemistry, coupled with concurrent conceptual reformulations, marched side by side with explorations into quantum theory, relativity, and the elucidation of the internally structured atom. The discovery of x-rays and spontaneous natural radioactivity, and the representation of the electron as a particle of discrete mass and charge, paved the way for the identification of α-, β- and γ-radiation, the establishment of the laws of radioactive transmutation, the nuclear theory of the atom based on α-scattering deflections, the cloud chamber, the isotope concept, and the planetary theory of the atom.

Escape into another world, after a sickening world war, led investigators to precision mass spectroscopy, α- and β-induced transmutations, measurement of the spin and magnetic moments and the diffraction of electrons, the theory of the α tunnel effect and nuclear potential barriers. It was suggested that the source of stellar energy might be nuclear fusion reactions; that the nucleus could be pictured as composed of protons and electrons with α-particles, composed of protons and electrons, as sub-units in the structure.

In his Bakerian Lecture to the British Association for the Advancement of Science in 1920 entitled "Nuclear Constitution of Atoms," Ernest Rutherford, with uncanny foresight - or so it seemed to everyone in 1932 - had postulated the possible existence and properties of a "neutral doublet." At that meeting he said: "On present views, the neutral hydrogen atom is regarded as a nucleus of unit charge with an electron attached at a distance, and the spectrum of hydrogen is ascribed to the movements of this distant electron. Under some conditions, however, it may be possible for an electron to combine much more closely with the H nucleus, forming a kind of neutral doublet. Such an atom would have very novel properties. Its external field would be practically zero, except very close to the nucleus, and in consequence it should be able to move freely through matter. Its presence would probably be difficult to detect by the spectroscope, and it may be impossible to contain it in a vessel. On the other hand, it should enter readily the structure of atoms, and may either unite with the nucleus or be disintegrated by its intense field, resulting in the escape of a charged H atom or an electron or both."

Before 1930 improved Geiger-Müller counters, electrostatic particle accelerators, the discovery of the Compton effect, the enunciation of the particle-wave concept, and matrix and wave mechanics gave scientists a bewilderingly novel and challenging array of phenomena and theoretical constructs to play with. In the laboratory, experimental design and instrumental ingenuity were stretched to the limit. At the blackboard and on paper, and with the help of the slide-rule, reckless, exciting, creative abandon was encouraged. Some physicists began to think about the feasibility of a quantum mechanics of the nucleus; but nothing could shore up the myth of simplicity that readily.

Niels Bohr, notably, came about as close to being au courant with what was going on in nuclear physics and chemistry as anybody on the scene in 1930. In his Faraday Lecture to the Fellows of the Chemical Society of London, on 8 May 1930, Bohr delivered a brilliant, if provocative and controversial discourse on the "Chemistry and the Quantum Theory of Atomic Constitution." In that account he reviewed the current outlook on the constitution of atomic nuclei, obviously stressing the importance of fundamental experimental discoveries and the accumulation of a great wealth of valuable information concerning the nucleus.

Against the background of all of the available information about the nucleus - radioactivity, α-scattering results, mass defect measurements, and the details associated with the sequential steps of known nuclear reactions - Bohr suggested that the most that could be agreed upon concerning the constitution of the nucleus, was that nuclear forces would have to be much stronger than the electromagnetic forces associated with atomic structure. Not even the nature of the forces that hold together the protons in the helium nucleus could be specified. Empirical evidence concerning charges and masses, as well as the known spontaneous and excited nuclear disintegrations, indicated however, as Bohr saw it, the plausibility of the assumption that all nuclei were built up of protons and electrons.

On passing from the helium nucleus to heavier nuclei, the problem of constitution became much more complicated. Nevertheless, Bohr indicated that "a certain simplification [the general facts of radioactivity, and small mass defects for helium] is afforded by the circumstance that the α-particle can be considered to a large extent to enter as separate entities into the constitution of these nuclei." However, any idea of constitution beyond that assumption, for even the simplest of nuclei, as far as Bohr was concerned, revealed that "the present formulation of quantum mechanics fails essentially." Recognizing at the same time, "the limitations imposed on classical electrodynamics" meant, of course, that Bohr's correspondence principle also was impotent.

Bohr had pinned his hopes for new insights and interpretations of what was going on in the nucleus, mainly, on the study of spontaneous and controlled nuclear reactions, but also on hyperfine spectra, and the magnetic and angular momenta of nuclei. In addition one would have access to nuclear statistics and probability considerations. Unfortunately the interpretation of such data also fell "largely outside the scope of present quantum mechanics." In fact, Bohr was not even convinced that the nucleus contained electrons; but since the negative electrification of the nucleus corresponded with whole numbers of elementary units, he felt that at least "the expulsion of a β-ray from a nucleus may be regarded as the creation of an electron as a mechanical entity."

The essential roadblock to more fundamental nuclear physics and to theoretical predictions of a more quantitative character, Bohr conjectured, was based on "ignorance of the forces acting on the α-particles and protons in the nucleus." Promise toward this objective, Bohr felt, would come from experimental investigations into controlled (i.e. induced) radioactive disintegrations, and more specifically, from x-ray expulsion studies. The crucial desideratum was to understand why the energies of the β-rays, unlike the α-rays, had been found to vary over such a wide and continuous range.

Might it be the case that β-ray expulsion process was not caused spontaneously by internal agency, but initiated externally? In the domain of non-radioactive elements no such external agency was known. Besides, the peculiar phenomena associated with β-decay were seen to be in no way a consequence of quantum mechanics. For want of any theory of intra-nuclear electrons, Bohr was led to entertain the idea of "a new fundamental feature of atomic stability." He concluded, that "at the present stage of atomic theory ... we have no argument, either empirical or theoretical, for upholding the energy [conservation] principle in the case of ß-ray disintegrations." Bohr concluded by saying that he would entertain no more speculations, and that he had "touched upon them here merely to emphasize that in atomic theory, notwithstanding all the recent progress, we must still be prepared for new surprises." A flood of surprises - an experimental *embarras de richesse* - was within earshot.

III. DISCOVERY OF THE NEUTRON

In October of 1930 Walther Bothe and Herbert Becker, at the Physikalische Technische Reichsanstalt in Berlin, announced in *Zeitschrift für Physik*, the startling discovery ("Kunstliche Erregung von Kern-γ-Strahlen") that the α-induced atomic disintegration of beryllium did not emit protons but gave rise to what appeared to be γ-radiation more penetrating than any heretofore encountered in radioactive disintegrations. This was contrary to what might have been expected, *viz.*, that this was another (α, p) reaction [α particles in /protons out] not unlike the reaction that Rutherford and Chadwick had discovered in 1919 when they demonstrated that: $N^{14} + He^4 \rightarrow O^{17} + H^1$

Although such (α, p) reactions with light elements were commonplace by the late 1920's - all α-induced disintegrations had been found to be (α, p) reactions - no investigator had encountered the level of energetic radiation observed by Bothe and Becker. The announcement intensified the already on-going spirited discussion about how nuclear emanation phenomena were to be monitored and interpreted from the data provided by cloud chamber and scintillation counter.

In a paper presented to the 18 January 1932 session of the Académie des Sciences, and published in *Comptes Rendus* ("Emission de protons de grande vitesse par les substances hydrogénées sous de l'influence des rayons α très pénétrants"), Irène Curie and Frédéric Joliot presented an account of their own version of the bombardment of beryllium with polonium α-particles. They had reached essentially the same conclusion about the production of an extremely penetrating radiation, apparently a γ-type emanation. In addition, however, they had observed that this emanation possessed the property of ejecting protons on passing through a paraffin block. They explained this unique result by suggesting that the γ-rays had produced recoil protons on the model that Compton had invoked for recoil electrons in 1923. The protons were seen to have velocities up to 3×10^9 cm/sec, corresponding to a quantum energy of the order of 50×10^6 eV.

At the end of February in 1932, a short note from James Chadwick at the Cavendish appeared in the "Letters to the Editor" column of *Nature* entitled "Possible Existence of a Neutron." Chadwick suggested that the penetrating beryllium radiation was the long sought-after neutron. He wrote: "Up to the present, all the evidence is in favour of the neutron, while the quantum hypothesis [viz. that γ-radiation accounts for the intensely penetrating radiation], can only be upheld if the conservation of energy and momentum be relinquished at some point." Three months later, in the *Proceedings of the Royal Society*, in a 17-page paper, entitled "The Existence of the Neutron" James Chadwick gave a more complete account of his own experimental investigations to support the "neutron hypothesis."

Chadwick recognized several difficulties with the energetics of the Joliot-Curie interpretation. He indicated that "the frequency of scattering of high energy quanta by electrons is given with fair accuracy by the Klein-Nishina formula, and this formula should also apply to the scattering of quanta by a proton .... [whereas] the observed frequency of the proton scattering is, however, many thousand times greater than that predicted by this formula." Secondly, Chadwick found it most difficult to account for the emission of a quantum of such high energy from the transformation $Be^9 + He^4 \rightarrow C^{13}$ + quantum.

Using a valve counter, Chadwick was able to demonstrate that similar effects could be produced for a number of light elements. Accordingly, he concluded that a stream "not of quanta but of particles of mass 1, and charge 0, or neutrons" had been liberated. Thus $Be^9 + He^4 \rightarrow C^{13} + n$. Knowing the masses of all of the isotopes in this reaction from the measurements of Francis Aston, and the experimental values for the kinetic energies of the particles, Chadwick calculated a neutron mass of 1.0067: "Such a value supports the view that the neutron is a combination of proton and electron, and gives for the binding energy of the particles about 1 to 2 $\times$ 10^6eV." This was reasonable.

The April 28, 1932 session of the Royal Society provided the occasion for re-examining questions about the structure of the nucleus that had been raised by the discovery of the neutron. At this meeting an approach to nuclear physics prevailed that was markedly different than what had seemed important at a previous meeting in February of 1929, on this same subject. All of the talks dealt in one way or another with the neutron. Lord Rutherford, again, presented the opening address on the subject: "Discussion on the Structure of Atomic Nuclei."

Rutherford reviewed the advances in mass spectrography, band spectra in relation to isotope ratios and nuclear spin, artificial transmutation, and α-scattering. More important for Rutherford as a guide to experimentation, was the "application of wave mechanics" to the explanation of the Geiger-Nuttall rule connecting velocity of escape in radioactive α-disintegration with the transformation constant. According to the theory (George Gamow, R. W. Gurney, E. U. Condon) the nucleus, composed of α-particles and other constituents, was surrounded by a high positive potential barrier that holds the constituents in equilibrium by strong attractive forces of a type unknown. On this theory, however, no detailed model of a heavy nucleus could be conceived that would hold together, in so concentrated a structure, particles as diverse as α-particles (mainly) with an admixture of free electrons and protons. With a nose for ferreting out the experimentally useful parts of a theory, and treating them as "suitable analogies" Rutherford focused on the implications of wave mechanics for α-particle expulsion. The essential point that Rutherford latched onto in Gamow's view was that the probability of α-particle escape, without loss of energy, would increase with the magnitude of the energy possessed by the individual α-particle in the nucleus. "This general conception of the nucleus," he remarked, "has proved very valuable in a number of directions and has been a very useful guide to the experimenter."

We may note here that much earlier in the century, the inherently complex energy spectrum of β-particles from radioactive substances, had been studied with considerable care in Berlin, at the Cavendish, at the Institut für Radiumforschung in Vienna, and elsewhere. Rutherford and his colleagues had been able to show that primary β-rays produced a continuous energy spectrum, which often was superimposed on discrete secondary β-ray spectra generated by conversions accompanying γ-rays. For many years this presented serious difficulties. An understanding of β-decay was so difficult simply because of the fact that, unlike the complexity of β-emissions, α-disintegrations from a single nucleus, with few exceptions, were seen to occur with nearly constant energy emission.

In October of 1931, at the nuclear physics conference in Rome, Niels Bohr in his Faraday Lecture on "Atomic Stability and Conservation Laws," stressed even more forcefully (than mentioned above in his lecture of 1930 at the Chemical Society) that nuclear disintegrations represented a departure from the law of conservation of energy. He maintained that: "Just as we have been forced to renounce the ideal of causality in the atomists interpretation of the ordinary physical and chemical properties of matter, we may be led to further renunciations in order to account for the instability of the atomic constituents themselves."

Eventually, as we know, energy conservation was salvaged by Wolfgang Pauli's neutrino hypothesis - which although conceived in 1930 and discussed by correspondence with Lise Meitner and Hans Geiger, was not published until after the Solvay Conference of 1933. Shortly thereafter Enrico Fermi incorporated the "little neutron" (Fermi coined the word "neutrino") into his elegant and influential theory of β-decay which was published in 1934 ("Versuch einer Theorie der β-Strahlen" *Zeitschrift für Physik*)

As we know from Rutherford's comments at the Royal Society in 1932, he was rather puzzled by the experimental complexity of the β-ray emissions. Convinced that all radioactive emissions, including β-rays would be better understood by studying all of the radioactive emissions simultaneously, Rutherford chose to focus on the more manageable α-rays and γ-rays. His point of departure was to assume that γ-rays from a disintegrating nucleus "represent in a sense the characteristic modes of vibration of the nuclear structure ... and that nuclear γ-rays are due to the transition of an α-particle between energy levels in an excited nucleus." He felt that the most profitable attack on nuclear structure would come from long range α-particle studies and the fine structure spectra encountered in β-particle emission. Essentially the experiments would give "direct information of the magnitude of a number of the possible energy levels in this particular [Radium C] nucleus." Rutherford's work with B. V. Bowen, in fact, had supported the general correctness of Gamow's theory (the relation of γ-rays to fine structure in the α-rays) by demonstrating that "the emission of α-rays is always accompanied by the appearance of γ-rays."

Any examination of Rutherford's priorities *à propos* nuclear physics at the Cavendish at this time reveals a comprehensive experimental programme designed to elucidate nuclear constitution and structure through the study of γ-rays: their origin and characteristic modes of vibration and excitation within the nucleus, the absorption of γ-rays (quantum energy) by different elements and the resultant nuclear excitations, the interpretation of γ-spectra from radioactive elements, and their relations to transitions in the α-particle between energy levels in the excited nucleus.

Given such objectives, Rutherford was motivated to pay serious attention to the frequencies that are defined by the quantum relations. He mentioned that "it has long been supposed that the quantum laws hold within the nucleus, and the correctness of this assumption has been abundantly verified in recent years." Clearly, Rutherford at this time was somewhat more enthusiastic about the relevance of quantum mechanics for the nucleus than Bohr.

Bohr's meandering explorations into all of the tributaries of both nuclear physics and quantum mechanics, had led him to say, only two years earlier, that quantum mechanics "fails essentially" on close inquiry "into the constitution of even the simplest nuclei." On the same occasion he had remarked that "the main sources of knowledge regarding the constitution of atomic nuclei," viz. "the study of their disintegration ... the hyperfine structure of spectral lines ... [that provide] magnetic moments and angular momenta of the atomic nuclei ... and the statistics obeyed by the nuclei," all lead to results whose interpretation "falls largely outside the scope of present quantum mechanics." Rutherford was on the lookout for clues to successful experimental investigations, even if the source was theory. Bohr's focus was more singularly and critically directed toward establishing the boundary conditions of quantum mechanics and the correspondence principle.

IV. HIGH SPEED ATOMS

In 1932 Rutherford, as the recognized architect of experimental nuclear physics, was 61 years of age. He still was in complete command of research style and priorities at the Cavendish Laboratory. He recognized that "speculative questions" were afloat – "like the possibility of the annihilation of matter and its conversion into radiation ... [and] the numerical relation between the unit of charge and Planck's constant h or ... the formation and transformation of nuclei, under the influence of conditions existing in hot stars." But rapid progress in the discipline, he felt, would come from other quarters, through new instrumentation that would produce "powerful but controllable high-speed atoms and high-frequency radiations to bombard matter." He specifically mentioned, in this context, the proton accelerating machine being constructed by his colleagues, John Cockcroft and E.T.S. Walton at the Cavendish, and the ingenious methods of high-speed atom production being developed by Merle Tuve, Lawrence Hafstad, and Odd Dahl at the Department of Terrestral Magnetism in Washington, and by Ernest Lawrence and Stanley Livingston at the University of California.

Advances in nuclear physics, in fact, were closely correlated with the accessibility of acceleration instrumentation. After 1932, the search for high-energy protons and deuterons developed into a race to achieve high voltages, and methods to control and amplify them by means of electrostatic generators, voltage multipliers, and high vacuum chambers. To avoid the technical limitations of high voltage breakdown, spiral "cyclotrons," using lower voltages were constructed in the early 1930's on the principle of the combined action of a homogeneous magnetic field and an oscillating electric field. By 1933, ions of 600 KeV were being produced for nuclear disintegration experiments by means of Van de Graaff electrostatic generators first built in Princeton. To accelerate electrons, that eventually overtook the cyclotron energy record, called for another technology. The first successful betatron not much bigger than a large suitcase, was constructed by Donald Kerst at the University of Illinois in 1940. It produced 2.3 MeV electrons and generated the energy equivalent of a gram of radium.

Even before Rutherford's Royal Society Lecture appeared in print, Cockcroft and Walton, after several years of high-voltage vacuum-tube preparations, had managed, with a voltage accelerator or multiplier tube, to accelerate protons to energies equivalent to 600 KeV (i.e., 600,000 volts) with their high-voltage accelerator at the Cavendish. On bombarding lithium with these swift protons they discovered that helium was produced: $Li^7 + H^1 \rightarrow He^4 + He^4$. This first artificial nuclear disintegration furnished the first experimental evidence for Einstein's mass-energy equivalence law.

V. EXPERIMENTAL NEUTRON PHYSICS

Nineteen thirty-two was the *annus mirabilis* of nuclear physics, for both experiment and theory. However, there was relatively little connection between what was being learned by those who cultivated an experimental approach to the discipline and what was conceived, formulated, and discussed at the level of nuclear theory. Chadwick's neutron discovery was followed by the spectroscopic and subsequently mass spectrographic identification of deuterium by Harold Urey, Ferdinand Brickwedde and George Murphy. The positive electron, or positron, or anti-electron, predicted by Paul Dirac in 1930, was identified in Wilson cloud chamber tracks in August of 1932 during the course of photographing cosmic-ray tracks. The following year, P.M.S. Blackett, G.D.S. Occhialini, Carl Anderson and Seth Neddermeyer were able to photograph electron-positron pair production.

By 1932, mainly through the work of Aston on improved mass spectrograph instrumentation, accurate atomic weights were available for some 250 isotopes. With this information it was possible to calculate the isotopic composition of elements, mass defects from mass numbers (usually interpreted on the model of a nucleus composed solely of protons and electrons), binding energies and packing fractions. The binding energy per proton turned out to be nearly constant for all elements, and was seen to follow from an almost constant packing fraction for all but the lightest elements.

Already in 1928, these observations were thought to have potential significance for explaining the differences between normal and anomalous (for hydrogen, magnesium, aluminum, and uranium) α-scattering as related either to electrical polarization forces, or conceivably, to the potential existence of stronger-than-Coulomb forces acting between nuclei. Normal "Coulomb-scattering" or "Rutherford-scattering" was used, for example, to refer to inverse square repulsive forces between two point charges such as the gold nucleus and the α-particle. The upshot of α-scattering experiments at the Cavendish (1919-1927) led to the conjecture that Coulomb repulsion broke down for the α-particle and the uranium nucleus at separations equal to the uranium nucleus, and that the α-particle might be captured by the potential barrier without subsequent α-decay.

In 1928 Gamow ("Zur Quantentheorie des Radioaktiven Kerns" *Zeitschrift für Physik*) and Condon and Gurney, independently, had suggested, that wave mechanics would allow for a finite probability that a particle could penetrate the potential nuclear barrier. Gamow, in 1929, used this idea in his model of the nucleus as a collection of α-particles interacting by strong attractive forces that fall off rapidly as a function of distance. Gamow conjectured that such an α-particle model of the nucleus would behave like a liquid drop held in shape by surface-tension-type forces. The idea, as reformulated by Bohr in 1936 in a number of published and unpublished papers and manuscripts, was invoked by Meitner and Frisch, to account for the observed phenomenon of nuclear fission in 1938.

In his Bakerian Lecture of May 1933, Chadwick, who had been on the look-out for Rutherford's neutral doublet for over a decade, gave an exhaustive account of experimental neutron physics at the Cavendish. The topics covered included: production of neutrons from boron and beryllium, neutron emission as a function of the range of incident α-particles, absorption characteristics of the neutron in air, mass of the neutron in relation to that of the proton, collision studies of neutrons with atomic nuclei, collision cross sections, disintegrations initiated by neutrons, and the production of positive electrons. This paper was loaded with new information, but rather slim on theory.

To return once more to the situation at the beginning of 1932, we may note that the immediate response in Paris to Chadwick's neutron hypothesis was one of incredulity followed by disappointment that a great discovery had slipped away. Undaunted, the Joliot-Curie team extended the α-irradiation investigation of nuclear reactions to other elements at the lower end of the periodic table like aluminum and boron.

In the session of 15 January 1934 of the Académie des Sciences, it was announced, in a report entitled "Un nouveau type de radioactivité," that the (α,n) reaction with aluminum generated a new, i.e. artificially induced, radioisotope of phosphorus which achieved stability by emission of "des électrons positifs":

$$^{27}_{13}\mathrm{Al} + {}^{4}_{2}\mathrm{He} = {}^{30}_{15}\mathrm{P} + {}^{1}_{0}\mathrm{n}$$
$$^{30}_{15}\mathrm{P} = {}^{30}_{14}\mathrm{S} + {}^{+}_{\varepsilon}$$

Although they had been anticipated in the discovery of the positron by Anderson et al., in 1932, they had synthesized the first new radioactive elements. They were quick to point out that projectiles other than α's should be explored to generate various artificially radioactive elements. In fact the stage had been nicely set for doing so with both protons and deuterons (discovered in 1932) using high voltage accelerators. Throughout the 1930's newly synthesized radio elements multiplied like mushrooms. In 1935 when Chadwick was awarded the Nobel Prize for physics, Irène Joliot-Curie and Frédéric Joliot jointly received the Nobel Prize for Chemistry.

VI. THEORETICAL NEUTRON PHYSICS

At the April 1932 Royal Society meetings referred to above, Rutherford's son-in-law, Ralph H. Fowler, who directed a theoretical group at the Cavendish, brought to discussion the puzzling problem of spin and statistics for the nitrogen-14 isotope. Nuclear scientists had known that they would have to come to terms with the N^{14} enigma. On the assumption that the nitrogen nucleus is composed of 14 protons and 7 electrons, and recognizing that an odd number of components should follow Fermi statistics with $\frac{1}{2}\hbar$ spin per particle, a composite angular momentum of half-odd integral multiples of $\hbar$ was to be expected.

In fact, by 1929 several experimental measurements had shown that nitrogen-14 had a spin of $\hbar$, thus following even-number Bose statistics. Almost immediately after the neutron was discovered, Iwanenko in Moscow, Chadwick in Cambridge, and Heisenberg in Leipzig, independently suggested that the N^{14} difficulties could be resolved by endowing the neutron with half-integral spins that followed Fermi statistics.

Elsewhere, across the Channel, the discovery of the neutron promptly gave rise to a great outpouring of theoretical discussions about nuclear forces and models of the nucleus. There notably were three physicists who attempted to construct comprehensive nuclear theories; their ideas on the subject have been argued about and built upon ever since: Werner Heisenberg in Leipzig, Ettore Majorana in Rome, and Eugene Wigner, an Hungarian at Princeton. All three adopted specific postulates regarding the nature of nuclear forces and nuclear structure based on the existence of the neutron. They then proceeded to draw consequences of the features of the postulated models of the nucleus. These ideas entered into the scientific literature as Heisenberg-, Wigner-, and Majorana-forces.

In June of 1932, five years after Heisenberg had managed to transform the mathematical formalism of quantum mechanics into a physical theory, - a move that put the underpinnings of large chunks of established theory in jeopardy - he published three classic papers: "Über den Bau der Atomkerne I, II and III" (*Zeitschrift für Physik*). It was another mathematical formalism, but this time invented to account for nuclear interaction forces and nuclear structure that could be described quantum mechanically. Protons and neutrons composed the entire nucleus, and were treated as distinct states - a notion that was amplified at a later date as the concept of isotopic spin.

The key idea for Heisenberg was to fall back upon an exchange interaction force between proton and neutron that was analogous to the idea of resonance interchange between a hydrogen atom and an H+ ion or proton, *viz.* the motion of one electron in the field of two hydrogen nuclei. This trick reminds one of the readily visualizable analogy that Rutherford had made use of in 1920 when he compared the interaction between the proton and the electron in a hydrogen atom with their interaction in the "neutral doublet," as a collapsed hydrogen atom. So, in a sense, Heisenberg was treating the neutron as a close coupling of a proton and an electron. Actually Heisenberg saw fit to postulate three kinds of forces acting within the nucleus: a strong force between proton and neutron, a weaker force between neutrons, and a Coulomb force between protons.

In his trilogy of theoretical papers, Heisenberg drew on the theories of Bohr, Pauli, Gamow, Slater and Placzek while ranging over a wide spectrum of potentially plausible ideas concerning the nucleus, its structure, and nuclear reactions. For example, he discussed: the case of proton-neutron interaction in which there is no contribution from electrons; the Hamiltonian model for the nucleus; the charge, mass, and especially the stability of the helium nucleus; stable and unstable (radioactive) isotopes; the stability of nuclei as a function of even and odd numbers of neutrons; the scattering of γ-rays from nuclei; Fermi's approach to interaction of nucleons; and the general nature of the properties and reactions of the neutron. However complicated Heisenberg's mathematical formalisms may have been, he had avoided some rather entangled experimental issues by backing away from the consideration of the physics of β-decay, and focusing instead on the motions and forces of protons and neutrons in the nucleus.

References to known experimental information about the nucleus were not the dominant theme in Heisenberg's papers. Apart from the fact that the neutron's existence was built into the core of the theory, mention was made in passing to such matters as: the degree of stability of nuclei in the radioactive decay series with respect to α- and β-decay; γ-scattering on nuclei; the Meitner-Hupfield effect; Aston's mass defect measurements in terms of binding energies; and the importance of investigating continuous β-spectra. The emphasis falls more heavily on the analysis of ordinary forces and exchange forces using quantum mechanical calculations, the statistics satisfied by β-decay theory, and the liquid drop model and barrier theory of Gamow.

It is appropriate to mention here that in 1933 Walter Elsasser in France and W. H. Guggenheimer in Germany took up the notion that nucleons could occupy quantum states. But this was an idea that already had been implicit in 1932, in the work of Bartlett at the University of Illinois, when he suggested that nucleons might occupy quantum groups or shells. Also, on the theoretical frontier in 1932, Dirac, Fock and Podolsky had proposed a quantum electrodynamical formulation to replace space configuration descriptions for charged particles. On the basis of the mass defect of helium and α-particle and deuteron binding energies, Wigner proposed short-range nuclear forces.

In 1933, in a direct and critical response to Heisenberg's paper, Wigner sought to explain the experimental observation that the binding energy of the α-particle is about 13 times as great as that of the deuteron ("On the Mass Defect of Helium" *Physical Review*). From a quantum mechanical computation for a nucleus composed of two protons and two neutrons, Wigner showed that "the difference of the mass defects of He and H^2 can be attributed to the great sensitivity of the total energy to a virtual increase of the potential – as is brought about by the fact that every particle in He is under the influence of two attracting particles, instead of one as in the case of H^2."

In Majorana's paper of the same year ("Über die Kerntheorie" *Zeitschrift für Physik*) an alternative to Heisenberg's proton-neutron model was proposed. This was structured upon a demonstration that a statistical analysis, in place of Heisenberg's Hamiltonian approach, would lead to nuclear exchange forces that avoided isobaric spin formalism. As Wigner had done, Majorana also dropped Heisenberg's neutron-neutron forces as well as the Coulomb repulsion for light atoms.

More important, Majorana rejected the analogy for the interchange of nuclear particles mentioned above. He did so, because Heisenberg's "theory does not explain the inner structure of the neutron, and the interaction between neutron and proton seem rather big compared with the mass-defect of the neutron as determined by Chadwick .... We shall use a statistical method ... and ... should like to point out [that] the exchange forces must have the opposite sign to Heisenberg's forces .... The symmetry characteristics of the eigenfunctions belonging to the normal state and the whole statistical treatment are different from Heisenberg's". Majorana also felt that the assumption of repulsive forces for such small distances in the nucleus were "aesthetically unsatisfactory," and that it was plausible "to try to find another solution and introduce as few arbitrary elements as possible."

Majorana's theory was much preferred by physicists in the 1930's, and it was not until some 20 years later that Heisenberg's views again came to be in vogue. In the meantime, in 1936, experimental evidence from proton-proton scatterings had shown that the nuclear forces were charge independent, and that they applied as well to the neutron-neutron and proton-proton forces as they did to the strong neutron-proton forces.

VII. NEUTRON IRRADIATION EXPERIMENTS IN ROME

Italian physics between the two World Wars was quickened by industrial, social and economic incentives spawned by the fascist regime. The Italian government encouraged university reform and scientific research by creating the Consiglio Nazionale delle Richerche. Enrico Fermi's patron, Orso Mario Corbino, the director of the physics laboratory at the University of Rome, was anxious to demonstrate that Italian science could be elevated to the level of theoretical physics that would meet European standards. Before the age of 30, Fermi had acquired an impressive reputation in theoretical physics. His contributions encompass the use of statistical mechanics in the enunciation of the ergodic hypothesis, tensor analysis applied to relativity theory, quantum theory,

theory of β-decay, and the statistical analysis of particles obeying Pauli's principle, i.e. Fermi statistics, or Fermi-Dirac statistics.

By 1934, three methods were available for probing the structure of the nucleus and initiating nuclear reactions: the use of natural projectiles emitted in radioactive decay; the acceleration of charged ions using electrostatic machines; and the irradiation of matter with the neutrons emitted from a radium-beryllium or polonium-beryllium source. In the case of neutron irradiations it was difficult to secure homogeneous projectile beams, and no acceleration of the particles is possible, but because of their electrical neutrality neutrons are able to penetrate deeply into the nucleus without loss of energy in ionizing the matter through which they pass.

By 1934 Fermi the theoretician had become convinced that progress in theoretical physics would depend primarily on new information acquired by experimental investigation. It was within this mental framework that he initiated a programme of team research in Rome focused on the systematic neutron-irradiation of all elements in the periodic table. Fermi was the boss, Emilio Segrè was responsible for securing pure elements, Franco Rasetti (who also acquired an international reputation as a paleobotanist) was put in charge of irradiation equipment, Edoardo Amaldi specialized in counters, Oscar D'Agostino (who had been trained in the Curie laboratory in Paris) served as expert chemist, and Giulio Cesare Trabacchi was responsible for securing the supply of radium. In March of 1934 this closely knit group, working on a single problem - probably the first such team research in the world - set out, with panache, on an experimental journey that would provide considerable competition for what was going on at the Cavendish in Cambridge, the Institut du Radium in Paris, and the Kaiser Wilhelm Institut in Berlin.

The Rome team began with experiments designed to determine whether bombardment with neutrons (from a radium-beryllium source) would produce phenomena of induced radioactivity similar to those observed with α-particles. Among the lightest neutron-irradiated target elements that were examined, using a thin aluminum wall Geiger-Muller counter that would allow β-rays to pass through, only aluminum and fluorine exhibited any effects. The pulse phenomena observed were explained as possibly resulting from an (n, α) reaction; e.g. $F^{19} + n^1 \rightarrow N^{16} + He^4$, followed by the β-emission transmutation of N^{16} into O^{16}. In a series of ten consecutively published papers entitled "Radioattivatà Indotta [alternatively Provocata] da Bombardamento di Neutroni," (*Ricerca Scientifica* 1934/5) Fermi and his group reported the results learned from bombarding some 60 pure chemical elements. Approximately half of these yielded nuclides never before known. They recognized after only a few months they they were confronted by complicated parent-daughter sequences, yielding nuclides with enormous variations in half-lives. It became a very messy subject, especially for physicists.

One of the most significant discoveries made early on, was that of anomalous intensities of activation, varying by a factor of a few hundreds, when a layer of paraffin and other hydrogenous substances were inserted between the neutron source and the target. The explanation offered was that neutrons rapidly lose their energy by repeated collisions with hydrogen nuclei, thus slowing them down and eventually reaching energies corresponding to thermal agitation.

A short paper by Fermi submitted to *Ricerca Scientifica* on June 6, 1934, was published concurrently, in a somewhat longer version (two-pages) in *Nature* (June 16, 1934) under the title: "Possible Production of Elements of Atomic Number Higher than 92." This announcement, subsequently blown up a bit by Corsino, and then featured in the newspapers, drew the attention of scientists and an Italian public eager to pounce on a phenomenal discovery made by their own countrymen.

The novel feature of this work, as Fermi reported it, was the recognition of a fundamental difference in the character of the induced radioactivity produced by n-irradiation of the elements, and the hitherto reported experiments with other projectiles. As Fermi put it: "the number of elements which can be activated either by the impact of an α-particle (Joliot) or a proton (Cockcroft, Gilbert, Walton) or a deuton (Crane, Lauritsen, Henderson, Livingston, Lawrence) is necessarily limited by the fact that only light elements can be disintegrated, owing to the Coulomb repulsion." For neutron bombardments, more efficient in producing disintegrations, all of the targets were seen to emit β-particles. This was to be understood, theoretically, on the model that the absorption by the nucleus of the bombarding neutron, produced an unstable nucleus with an excess of neutrons. The re-establishment of stability was accomplished by the transformation of a neutron into a proton and an electron.

The evidence from the chemical separations, shapeless though they were, seemed to show that "three main processes were possible: (a) capture of a neutron with instantaneous emission of an α-particle; (b) capture of the neutron with emission of a proton; and (c) capture of the neutron with emission of a γ-quantum, to get rid of the surplus energy." The (n, α) and (n, ρ) reactions mentioned here, had been treated at some length in the previous literature. It was the n-capture possibility that created all the fuss - at the time, and almost continuously from then on until nuclear fission was discovered in Berlin in December of 1938.

Fermi saw immediately that "it seemed worthwhile to direct particular attention to the heavy radioactive elements thorium and uranium, as the general instability of nuclei in this range of atomic weight might give rise to successive transformations." The very large uncertainties in the decay curves due to statistical fluctuations revealed that the phenomenon was so complex, especially for uranium, that successive and alternative disintegration processes could not be readily established.

The chemistry of these heaviest elements turned out to be treacherous, but Fermi and his colleagues felt semi-confident that the "13 min-activity" product from n-irradiated uranium could not be any of the isotopes from uranium 92 down to lead 82. Fermi wrote: "this negative evidence about the identity of the 13 min-activity from a large number of heavy elements suggests the possibility that the atomic number of the element may be greater than 92." He concluded cautiously: "the possibility of atomic number 94 or 95 is not easy to distinguish from the former, as the chemical properties are probably rather similar .... It seems therefore at present premature to form any definite hypothesis on the chain of disintegrations involved." Nevertheless, the idea of trans-uranic elements was out; it occupied center-stage for the next four years in those scientific enclaves where experimental nuclear physics was fostered.

By the beginning of 1935, there appeared a grand summary article, authored by Amaldi, D'Agostino, Fermi, Bruno Pontecorvo, Rasetti, and Segrè; it was communicated to the Royal Society by Lord Rutherford. The characteristic features of all that had been learned by the Rome group was here in place: the methods adopted for isotope separations and identifications; the cross-section capture energies; the scattering, diffusion and resonance characteristics of neutrons as a function of speed (energy of the neutrons seen as the crucial parameter); the major differences between light and heavy elements; the production of new isotopes that were either one or two ordinal numbers below the target element or else where isotopes of the target element, and the most revolutionary but somewhat ambiguous case of n-capture that produces the trans-uranic elements.

Recall that Fermi, who was 34 years of age of 1935, already some years earlier had earned something of an international reputation as a theoretical physicist. Rutherford's gut feelings about the relative merits of experimental physics are admirably revealed in the first letter he ever wrote to Fermi dated 23rd April 1934.

He thanked Fermi for sending him an account of his neutron experiments, and hoped for "more information as to the actual mechanism of such transformations," it being "by no means clear that in all cases the process is as simple as appears to be in the observations of the Joliots." Rutherford concluded: "I congratulate you on your successful escape from the sphere of theoretical physics! You seem to have struck a good line to start with. You may be interested to know that Professor Dirac also is doing some experiments. This seems to be a good augury for the future of theoretical physics!" With such warm blessings emanating from the Cavendish, Fermi was given a personal entree into the universe of the experimentalists. Indeed, during the remaining two decades of his career, and especially in connection with his contributions to the "Manhattan Project," physicists throughout the world came to know Fermi as one of the very few physicists of this century who had reached his level of excellence in both experiment and theory.

By 1935, experiments on the slow neutron-irradiation of the heaviest elements were taken up in Berlin, Paris, Vienna, Copenhagen and New York. The chemical complications associated with the unambiguous establishment of the products of the reactions mushroomed so rapidly that Fermi and his crew abandoned the whole business. Especially after the implementation of Mussolini's racial laws (Manifesto della Razza, July 14, 1938) Fermi's group began to disperse. Those that remained, opted to focus on the less-explored, and more manageable elements in the middle of the periodic table. In any case, the existence of transuranics seemed securely anchored, even if the chemistry was horribly laborious and entangled. Fermi knew that the world's most prominent radiochemists had backed up the transuranic claims made in Rome; moreover that these chemists were far better equipped than physicists to straighten out the details in the multiple-branched parent-daughter sequence relations of the radioactive products.

An historical investigation of the experimental investigations that led to the discovery of nuclear fission at the end of 1939, leads through a labyrinth of chemical and physical maneuvers that is far more interwoven than anything discussed thus far in this paper. The unusual behaviour of n-irradiated uranium was resolved eventually only by putting into practice an entire armoury of novel intellectual and physical tactics: micro-analytical and carrier techniques, trace-product identifications and classifications, preparation of very pure samples, characterization of nuclear isomers, meta-stable and excited-state isotopes, resonance absorption and capture, establishment of extremely short half-lives, manipulation of variable speed neutron sources, and the counting, screening and monitoring of radiations. The subject merits a full-scale historical analysis and interpretation, but not on this occasion.

VIII. CONCLUDING REMARKS

Max Born, in his little book on *Experiment and Theory in Physics* (1944) sought, as he says, "to show in a simple way .... the mutual relationship between theory and experiment in the actual historical development of science, and to offer a balanced opinion on the present situation and future possibilities." In reference to developments such as the symmetry properties of wave functions and Pauli's principles, Born was convinced that they were derived "by a long inductive process, in which flashes of imagination have alternated with painful observation and interpretation of facts." This represents the "ideal cooperation of experiment and theory," such that there is "neither the experimenter boasting about the empirical purity of .... results, nor the theorist claiming *a priori* knowledge, but mutual help and encouragement."

This ideal two-way reciprocity between experiment and theory has not always been feasible. At least, it seemingly was not so in regard to the state of maturity that nuclear physics had achieved in the 1930's. In his account of the historical development of this inter-relationship, Born asserts: "From about 1930 experiment was ahead of theory." The two outstanding theoretical predictions mentioned by Born are the positron and the meson; but he adds: "However admirable, these discoveries are not products of pure reason, but the final outcome of a long chain of empirical research." Born concludes by saying: "My advice to those who wish to learn the art of scientific prophecy is not to rely on abstract reason, but to decipher the secret language of Nature from Nature's documents, the facts of experience." Thus speaks Born, the theoretician.

In my opinion, the history of nuclear physics, in general, provides a paradigmatic example of this situation vis-a-vis the reciprocity that develops between experiment and theory. It is not so much that nothing worthwhile is going on in theoretical physics, quite to the contrary. It is rather that the theoretical advances were not exerting a great deal of influence on what the experimentalists were engaged in exploring. The wave of new experimental findings initiated by the discovery of the neutron, including nuclear fission, self-sustaining chain reactions, etc., were surprisingly far removed from anything that had to do with quantum mechanics, the electromagnetic properties of particles, or models of the nucleus.

As one looks back, it is easy to identify important components of the total picture, here and there. But nuclear physics in the 1930's - a happy time for physics because so much was happening - appeared to experimentalists and theoreticians alike as somehow not moving in the same direction. Surely there were breakthroughs on the experimental side, and especially in the instrumental advances that fed those experiments. It was not at all difficult to keep surprising the theoreticians, but often the experiments were seen to be trivial, or else they turned out to be simply wrong. Phenomena that were on hand, of course, were picked up later and recognized to be important. That too is a part of the historical record, but clearly to be recognized as scientific hindsight.

In trying to establish a roster of fundamental steps in the genesis of nuclear physics, the most conspicuous observation to make, it seems, is the almost total omission of reference to theoretical papers on the part of investigators preoccupied with experimental nuclear phenomena. The theoreticians, by contrast, were swimming rather aimlessly in a sea of novel experimental phenomena about the nucleus that did not fit readily with any of their theoretical moves. There was an ever growing mass of experimental data, but very little that seemed in any way fundamental for the theoretician. Perhaps even the experimentalists - there actually were not very many investigators engaged in nuclear physics in the 1930's - were not having too much fun either. This state of affairs might well be chalked up to inherent complexities with nuclear phenomena, but also to a situation in which the theoreticians either were too confused to contribute something relevant to the experimentalist's craft, or too remote from the world of the laboratory, or both.

CHRISTOPH MEINEL

EMPIRICAL SUPPORT FOR THE CORPUSCULAR THEORY IN THE SEVENTEENTH CENTURY

1. INTRODUCTION

The purpose of this paper is to evaluate the empirical content and experimental foundation of the early seventeenth-century corpuscular hypothesis. Supported by ontological, epistemological, and mathematical arguments, observational and experimental evidence played an important part in creating those patterns of thought that were instrumental in the transformation and, eventually, acceptance of the corpuscular theory of matter.

Compared to the rise of astronomy and mechanics, the success of seventeenth-century atomism is ambiguous. Unlike other physical theories of the time the atomic doctrine was not based on experiments likely to be accepted by today's scientific standards. In Galileo's inclined plane and his law of falling bodies or in Newton's theory of colors and his *experimentum crucis* with the prism, for example, theory and experiment, observation and conclusion were connected in a way still acceptable to us. In atomism, however, this is not the case, although almost all corpuscular theories of the seventeenth century explicitly claimed to be derived from and based upon experience. Yet, it was not until the nineteenth century that experimental results made the atoms at least plausible.

The difficult relationship between seventeenth century atomism and its experimental foundation has been obscured to some extent by later historians. When the standard histories of atomism were written at the end of the nineteenth century[1], a final experimental confirmation of the corpuscular nature of matter was still lacking. Twentieth-century historians, on the other hand, have underestimated the empirical difficulties involved in early modern atomism.[2] Knowing that there are atoms in nature, they could hardly imagine how the protagonists of the scientific revolution should not have arrived at this same conclusion.

D. Batens and J. P. van Bendegem (eds.), Theory and Experiment, 77–92.

2. EXPERIENCE AND EXPERIMENTS

The empirical arguments presented to support atomism during the first half of the seventeenth century were mainly based upon five groups of phenomena:

1. Extrapolations from macroscopic bodies;
2. observations by means of the microscope;
3. transport processes such as evaporation, abrasion or growth;
4. condensation and rarefaction including the question of the vacuum;
5. chemical processes such as reductions of metals.

2.1. *Extrapolations*

Traditional arguments put forward - for the first time in antiquity and most of them in Lucretius' *De rerum natura* - were extrapolations. The most frequently quoted example was the one of insects which are so small that their third part would already be beyond the limits of visibility. How small then, Lucretius asked, must their organs be, and how small, in the end, the atoms out of which everything is composed?[3]

Originally aimed at giving but an idea of atomic dimensions, the observation was quoted frequently in order to prove their very existence. On a similar level lies the comparison of the atoms to the size of tiny motes one sees dancing back and forth when a ray of sunlight falls in a dark room.[4] In either case the empirical facts referred to were *loci classici* from a literary tradition of figurative use of pictures, aimed at creating astonishment and, through astonishment, assent and persuasion.

Authors scarcely expanded on these examples; they simply quoted them. In 1643 Claude Bérigard presented a carefully designed experimental verification of Lucretius' motes in the sunbeam.[5] To exclude the possibility that the phenomenon was caused by major particles such as normal dust, he sealed a glass vessel and kept it quiet for a long time making sure that all dust had settled. The minute reflecting particles he observed nevertheless inside the glass were consequently judged to be the atoms themselves.

Among those to make the first, if cautious, steps towards a quantitative determination of atomic dimensions was Daniel Sennert in 1636. In a series of experiments[6] designed to prove the existence of atoms Sennert described a distillation in which a stream of alcohol vapor passed through a sheet of paper, the density of which was supposed to give an idea of how small the atoms really were. This certainly was an impressive experiment, but at the same time a tacit reference to Lucretius.[7] Sennert went on to give examples of various distillations, comparing the enormous volume of vapor and the myriads of atoms in it to the small droplets into which they condense, or the smoky wick of an extinguished candle to the huge volume of air that was filled by its smoke.

The language in which these observations were described abounds with quantitative statements such as to the duration of the experiment, the number of corpuscles, the amount of the product, and the size of the candle-wick. However, it is clear that there was no quantitative methodology behind these indications. Sennert used the language of the laboratory in a merely figurative and persuasive manner, appealing to the imagination of the reader.

On the basis of impressive calculations Johann Magnenus even suggested exact figures for the size of an atom.[8] But what does it prove for the result computed in such a way to come surprisingly close to modern figures? Were such calculations really quantitative science or merely a scholarly variation of a theme from Archimedes' *The Sand-Reckoner*? Playing with numbers seemed quite common, indeed. In 1654 Walter Charleton published exactly the same calculation[9], though at least he should have been aware of the warnings by his admired hero Gassendi: Referring to Archimedes' attempts to compute the number of sand grains that fit into a poppy-seed, the French philosopher had already pinpointed the methodological problems involved in transferring this kind of geometric reasoning to physical matters.[10]

However, since Magnenus' calculations have been recently called the beginning of the quantitative methodology in atomic physics[11], it might be worth while examining more closely the attitude of this allegedly scientific mind towards experience and experiment. There is no doubt that Magnenus favored the empirical and mechanical spirit of his age; nevertheless his work abounds with purely dialectical reasoning and syllogistic conclusions. The only 'real' experiment he presented in great detail was taken from a popular compendium of natural magic[12]: According to Quercetanus[13] a Polish physician was reported to keep a collection of sealed glass vessels containing finely ground flowers of various kinds. When a candle was put underneath the vessel the corpuscles coalesced under the influence of heat to form a perfect blossoming plant. When the candle was removed, it again disintegrated. This strange experiment acquired some fame among the attempts of proving the reality of the atoms and was discussed even among the most respectable scientists of that time.[14] Needless to say, this was not the kind of experiment upon which a scientific atomism could have been properly established.

2.2. *Microscopical Perception*

Seventeenth-century science was fond of the small, the worlds to be found in a drop of water. There was a widespread enthusiasm for the magnifying glass and for the microscope which had just been invented. The new instruments made it possible to come closer to the details, closer to reality, and - so it was assumed - closer to truth. The possibilities of optical ingenuity seemed unlimited. In a chapter on the size of the atoms Gassendi thoughtfully meditated about what degree of refinement the borderline between man's and nature's subtlety might be extended to by the use of the microscope.[15]

What had been a merely potential aptitude of the instrument to Gassendi, was presented as an empirical fact by Henry Power a few years later. His *Experimental Philosophy*, the first English book on microscopy, claimed enthusiastically that the microscope enabled men to "see what the illustrious wits of the Atomical and Corpuscularian Philosophers durst but imagine, even the very Atoms and their reputed Indivisibles and least realities of Matter."[16] This was more than the selling rhetoric so common with prefaces. Power studied traces of mercury and found that the "atoms of Quick-silver [...] seemed like a globular Looking-glass"[17]. From the heterogeneity and particulate structure of a cosmetical precipitate, he inferred that "all the globular Atoms of current and quick [mercury]; besprinkled all amongst those Powders, like so many little Stars in the Firmament" remain unaltered when a compound is formed and retain their true nature.[18]

As a serious scientist, however, Power had to admit that he failed to succeed in seeing any corporeal effluvia by means of his optical device, although others had claimed to have seen the magnetic effluvium. Power, the meticulous observer, was convinced that such an observation "indeed would be an incomparable Eviction of the Corporeity of [...] Effluviums, and sensibly decide the Controversy 'twixt the Peripatetic and Atomical Philosophers"[19].

Stimulating as these instruments were for the study of biology, their meaning remained ambiguous in matter theory. Joachim Jungius used the magnifying glass for a thorough study of apparently homogeneous substances. He observed that they were in fact always heterogeneous if viewed through a microscope, and no surface could be so smooth that one could not think of a more powerful microscope that would reveal its real discontinuity. Consequently, Jungius stated that continuity must be exempt from the jurisdiction of sensuous experience. On the other hand, if there were no truly continuous parts in the end, endless progression and divisibility would result.[20] This was the vicious circle of every empirical approach to the atoms. Since they are so small that they could be inferred only rationally, methodological difficulties and a contradictory epistemology arose if one attempted to model the real after the visible.

2.3. *Material Transport*

Transport phenomena in which material substances appear or disappear invisibly provided a similar type of argument: Smell is an efflux of corporeal particles and comes to an end once its material source is exhausted.[21] Clothes hung near the seashore become wet and dry again in the sunshine.[22] The drying of bread or the slow evaporation of liquids are material processes, although the flux of material cannot be observed. In all these cases, quantitative change can be recorded, and from this the existence of invisible parts of matter could be inferred. Yet, this was not a very original conclusion, but rather an application of the theory of effluxion proposed by Empedocles, empirically founded by Democritus and Asclepiades of Prusa and eventually collected in Lucretius' poem. In the seventeenth century these examples were repeated over and over again, and similar ones were added.

It is hard to believe that this kind of naive realism was taken more seriously than we can take it with regard to the question of atoms. Three aspects should be given particular consideration: the phenomena dealt with so far suggested that the ultimate constituents of matter were (i) potentially observable by extended experimental effort, (ii) deductible by analogy, and (iii) provable by virtue of their actions. Yet there is little doubt that they were not suitable to definitively decide the question of matter, and they were certainly not understood in this manner by the contemporaries. The frequent occurrence and repetition of these observations, however, the persuasive idea that truth should be visible or could be thought of in a pictorial way infiltrated the scholarly discourse and the very language of science. In the end the idea of little particles as constituent parts of bodies became plausible, even though philosophically not entirely legitimate. Furthermore, and this was of no less importance, it transferred the corpuscular question to another level of inquiry. For now it became necessary to prove the material identity of these particles. This was, of course, a chemical question, but we have to deal with the more physical arguments first before we return to chemistry.

2.4. *Condensation and Rarefaction*

Unlike the instances of naive induction from sense perception dealt with so far, condensation and rarefaction and, above all, the vacuum, belong to a field of physical experimentation where one would expect more convincing departures from the traditional ways of reasoning. In ancient atomism, with its hard and impenetrable atoms, change required motion, and motion required a void space to move into. The nature of this void, however, was a matter of endless controversies until the eighteenth century.[23] Two alternatives had already been discussed by the Greeks: the continuous or three-dimensionally extended void and, secondly, the more widely received idea of a discontinuous, interspersed void between the particles of matter.

There was one classical experiment that used to be interpreted as evidence supporting the existence of microvacua: A vessel filled with loose ashes holds as much water as the empty vessel, because the tiny ash particles are received within the pores or vacua of the water. The observation was reported by Aristotle[24], referring to those who believed in the void, but was rejected on the ground that two bodies cannot occupy the same space simultaneously. This was indeed puzzling and kept the medieval commentators busy. Francis Bacon was presumably the first to disprove the phenomenon in question experimentally.[25]

In reality, things were not as simple. In Gassendi's *Syntagma philosophicum* the ash experiment was rejected on both experimental and philosophical grounds.[26] Instead, the French philosopher proposed another and more convincing experiment. He took water, saturated with ordinary salt, and found that this solution was still as capable of dissolving alum as pure water would have been.[27] From this Gassendi concluded that there must be various differently shaped microvacua in the water, each kind of which receives exactly one kind of corpuscles, e.g. a cubic space a cubic corpuscle such as salt, and an octaedric space an octaedric one such as alum. But wasn't it absurd to assume that the volume remained constant when another volume was added? This is exactly what Gassendi's carping critic Jean-Baptiste Morin thought. He repeated the experiment more carefully in a glass flask with a graduated neck and found that when salt, alum, and sugar were added to water, the volume of the resulting solution was greater than that of pure water.[28] From textual evidence alone, it is difficult to judge who was correct. In fact, either observation may have been correct: There are indeed certain salts that do not increase the volume of water when they are dissolved, and water-free alum is one of them.

Apart from explaining solution, the hypothesis of interparticulate vacua seemed especially helpful in understanding coherence. The standard experiment was the separation of two entirely flat surfaces from direct contact in order to show that during this process a void must result since the air cannot fill the entire space opened instantaneously.[29] Though originally supposed to prove the existence of a vacuum, the experiment soon acquired a crucial position among the proofs for its non-existence. It was still in this sense that Galileo, who rejected the extended void, referred to it as a perfect illustration for nature's abhorrescence of a vacuum.[30] He described a hydrostatic experiment designed to measure the breaking force of a water column that would give him a quantitative value for what he called *la resistenza del vacuo*.[31] It was exactly this "resistance of the vacuum" which he believed to be responsible for the strength and rupture tension of solid bodies.

The problem of coherence and the void became even more acute in condensation and rarefaction. The Aristotelian explanation was, that a given amount of matter could assume, at different times, contrary qualities; and since dense and rare were contrary qualities, the same amount of matter could occupy different volumes at different times while maintaining its identity through the changing modes of its existence.[32] Indeed no corpuscular explanation of a similar smoothness was available. Few authors would have admitted that the interspersed vacua could be blown up to a size that would account for the observed change of volume during evaporation. Otherwise, they would have had to admit a continuous vacuum. It is well known that, even after the Torricellian vacuum had been experimentally demonstrated in 1643, it was by no means unanimously considered to be entirely void. The first atomist to discuss the Torricellian experiment in great depth was presumably Gassendi; but even his account on the vacuum remained vague as to its meaning in theory of matter.[33]

Instead of admitting the void, most authors assumed some kind of ether or spirit that filled the spaces between the vapor atoms and glued the corpuscles of solid bodies together. Yet, the reintroduction of an active spirit or ether into atomism, aimed at explaining how the atoms interact and how their actions are transmitted, undermined the theoretical consistency of the mechanical corpuscularianism, while the question of change and motion remained, however relegated to a somewhat lower level. Plenist corpuscular theories such as Descartes'[34] exemplify that it was entirely acceptable to assume corpuscles without admitting the void.

2.5. *Chemical Change*

The chemical arguments in support of the atomic view of matter refer to two kinds of phenomena: Processes generating a new mixtum, or processes recovering constituents of a compound in their former state. In both cases it was to be explained how distinct atomic particles interact and how, from this interaction, new qualities emerge that were not originally present in the reactants. The emergence or eduction of a new form during substantial alterations was indeed the great theme of early seventeenth-century peripatetic theory of matter.[35] Here purely mechanical action, and especially local motion, was not a sufficient explanation. Therefore additional hypotheses came into play which did not originally belong to, or were even contradictory to, the principles of atomism, such as the corporeal ether, active spirits, neoplatonic concepts of sympathy and antipathy, or the teleology of directing forms acting upon moving particles.

On the other hand, there was the old tradition of metallurgists and practical chemists who ignored such abstract questions and naively took the original reactants and their properties as the true constituents of a compound.[36] It is significant that Sennert explicitly admitted that there was no way of proving the mechanism by which the unity of parts was effectuated and the form of the new compound generated.[37] Practical chemist as he was, Sennert preferred to leave such questions to others, and went on to argue that at least one thing was certain, namely that every mixture could be resolved into those parts of which it was originally constituted. For, if the identity of the constituent parts was not constant, there would be a generation of new constituents during the process of resolution and decay.

Of course, this was a tautology and depended on how one defines 'constituent part'. Yet, it was an important step within the 'chemical' argumentation in favour of corpuscles. For now, it was no longer necessary to bother with substantial change and the emergence of new qualities; instead, the question was reduced to a test of identity in a cyclic process. Once the identity of the original reactants and the final product could be demonstrated experimentally, the persistency of substantial identity in a material carrier would be proved, no matter how many alterations had occurred between the starting and final substances. This was clearly a departure from the former preoccupation with the quiddity of processes, and from the ontological level of the atomic theory to something that might be called a 'black box theory' of chemical change.

2.5.1. *Reduction*

There is ample evidence that the ground for this new perspective had been laid by the pragmatism and the atheoretical attitude of metallurgists and iatrochemists. However, such people were hardly literate and rarely uttered their theoretical assumptions. Angelus Sala, court physician, pharmacist, and advisor on commercial subjects, described pragmatically and to some extent also quantitatively the formation and decomposition of copper vitriol.[38] In doing so he distinguished, as usual, between transmutations and changes that resulted from a mere juxtaposition of particles. These latter processes were either *coniunctiones*, such as the alloying of gold and silver, or *reductiones* by means of which little particles were reassembled into their former coherent state.[39] The best example was a solution of gold in *aqua regia* and its precipitation using metallic silver, a process which Sala considered a mere division and rearrangement of metal atoms. He believed the metal to retain its substantial identity, though "hidden" because of its dispersion into single atoms. As a practitioner Sala did not bother with the nature of this "hiding" of qualities; instead he accepted the reduction to the pristine metallic state as sufficient proof of the identity of gold throughout this process.

Daniel Sennert, on the other hand, the learned professor of medicine, preferred to maintain the teleology of directing forms to account for the specific properties. Expanding on Sala's approach he conceived a more convincing argumentation, based upon a great variety of processes which he classified as reductions to the pristine state.[40] The first type were simple distillations and sublimations of substances such as alcohol, sulphuric acid, and sulphur, which he regarded to be merely mechanical operations by means of which bodies were mashed into their atoms.

A chemically more sophisticated argument came from the reduction of different mercury compounds to running *mercurium vivum*. In order to prove that, indeed, real atoms of mercury were involved and not just new substantial forms generated, Sennert referred to the nightmares of Paracelsian medication: Ointments and fumigations with mercury had the effect that a coin, put in the patient's mouth, became amalgamated - not to mention the findings of a postmortem autopsy. In all these examples the reduction to the pristine state was the decisive criterion.

The same is true for another type of experiment presented to prove that it was not the substantial form of a mixture that preserved the identity of its compounds, but in fact the atoms themselves[41]: Sennert fused gold and silver together to obtain an entirely homogeneous alloy. Then he poured aqua fortis or nitric acid on it. The silver was dissolved, whereas the gold particles settled to the bottom. He separated the two phases and precipitated the silver from the solution to obtain another fine sediment. Eventually, he melted both powders and obtained, quantitatively, gold in the first case, silver in the latter.

All these experiments were of course not 'invented' by Sennert, and it is irrelevant whether he actually performed them. Yet, he was the first to connect them systematically in order to demonstrate that the *reductio in pristinum statum* could be used as an argument in favour of atomism. It is interesting to see that Sennert, coming to this conclusion in 1636, had to revise his former ideas concerning the transmutation of metals from the corpuscular point of view.

2.5.2. *Transmutation*

Metallic transmutation was supported by old and well documented evidence, suitable as a defense of both the Aristotelian doctrine of substantial alteration and the alchemists' quest for gold: Certain springs and rivers had the peculiar property that a piece of iron, upon being immersed for some time, turned into true copper, first at the surface, and later throughout. This process, called cementation, had been known for a very long time and had been described by scientists whose credibility was beyond any doubt.[42]

Few authors denied these observations, among them the Lorraine physician Nicolas Guibert, whose explanations were ridiculed by Sennert in 1629. This was not just because Sennert at that time was still an Aristotelian, but because he could rely on experimental proofs.[43] However, when, in 1651, the posthumous edition of Sennert's collected works was published, the editor had to add a note found among Sennert's manuscript remains.[44] In it Sennert admitted that the alleged transmutation was presumably a mere separation of copper from its vitriolic solution by means of iron, and correctly considered this process as an exchange of atoms that retain their chemical identity.

Sennert went even one step further and conceived an experiment by which the truly quantitative character of the cementation process was to be proved: One ounce of copper was dissolved in sulphuric acid to obtain the acid, blue solution of copper vitriol. Then two ounces of iron filings were added. The vessel was kept in a warm place until all of the iron had disappeared. As a result the blue colour of the solution turned colorless, and a red precipitate settled at the bottom. This precipitate was washed, dried, and reduced, to yield exactly one ounce of copper – the very same amount that had been dissolved originally.[45] Again the method of *reductio in pristinum statum* was applied to demonstrate the existence of atoms. In doing so, Sennert was able to avoid the difficulties involved in explaining what happened to the properties of the metal when it 'united' with the 'salt' in solution, or whether the particles of copper, salt, and water form an essentially uniform mixture or a mere juxtaposition of parts.

Taken together, the chemical evidence presented by various authors to support the corpuscular view of matter supplied good empirical reasons for regarding natural bodies as divisible into much smaller ones that somehow retain the specific properties and can frequently be recombined to yield the original body. Yet the range and depth of experimental proofs for the atoms as such remained limited. A limited number of standard experiments was referred to more frequently in writing than repeated or extended in the laboratory. Attempts to widen the experimental basis of the corpuscular theory of matter were exceptional and barely convincing.

3. CONCLUSION

It is difficult to believe that arguments of this kind were capable to convince those who did not already share the atomic view of matter. Writing in the 1660's even Robert Boyle had to concede that the corpuscular hypothesis had not yet been, and probably never could be proved conclusively.[46] Mere extrapolations from the visible to an underlying invisible reality were philosophically naive and epistemologically questionable. Distillations, evaporations, growth of crystals etc., were certainly appropriate for showing that something material was transferred from one place to another, but they did not prove its corpuscular nature. Although the atoms would have supplied an easy mechanism, the corpuscular interpretation of rarefaction and condensation remained questionable, since the entire problem of how the atoms interact and cohere was open, and the existence of an extended void would have had to be admitted. The introduction of a material ether disposed of these problems while creating new ones by the strange hybridization of particulate and continuous matter. Thus, as far as the experimental support for the atomic theory was concerned, the chemists and iatrochemists offered the more convincing, though by no means decisive, arguments.

By its very nature the chemical approach was pragmatic, realistic, and eclectic. The majority of chemists worked on real matter and real properties in a purposeful way. After all, they wanted to sell a product or to cure a patient. They simply could not afford to rely too closely upon a rigid theory, not even a corpuscular philosophy of the Boylean kind. Needless to recall that, as opposed to a stubborn historiographical myth, Boyle's clock-work universe, which provided so many fertile concepts to seventeenth-century physics, "proved a sterile and occasionally adverse intellectual climate for an understanding of the processes underlying chemical change."[47] It was not the mechanical philosophy that was to succeed in chemistry, but a non-committal, substance-oriented notion of corpuscle, something that was more like an elementary particle or a small amount of substance than a Democritean atom.

During the subsequent development of chemistry Baconian empiricism and the naive realism of the chemists were to prevail over the philosophically consistent atomism of the late seventeenth century. By the standards of Boyle's corpuscular philosophy and John Locke's insistence on the epistemological status of the corpuscles[48], there was nothing of its kind throughout the eighteenth century. The new interest focussed on elements and affinity, not on atoms and motion.[49] Its non-committal character was certainly not the least advantage of the kind of corpuscularianism favoured by the 'chemical' atomists. The restricted nature of its epistemological support enabled the resulting notion of corpuscle to assume whatever requirements future research was to find convenient. The traditional insistence on the ontological conformity of science and nature, the requirement of proof or falsification by means of experiment, and, what is more, the very question of the truth-value was dismissed in favour of a merely instrumental or operational link between theory and reality.

In 1661 Henry Power, the microscopist, still believed that scientific knowledge, with the help of the new instruments, would eventually advance to the very atoms and least realities of matter. "A part of Philosophy," he wrote, "but yet in discovery; and will, I fear, prove the last Leaf to be turned over in the Book of Nature."[50] To Robert Boyle, on the other hand, whose reading experience with that book was derived from 'chemical,' not microscopical observations, it was of less concern which key was the true one to decipher nature's cryptograms: Any key would do provided it made coherent sense and was fit to solve the phenomena, for which it was devised:

> In the physical explications of the parts and system of the world, methinks, there is somewhat like what happens, when men conjecturally frame several keys to enable us to understand a letter written in cyphers. For though one man by his

> sagacity have found out the right key, it will be very difficult for him, to prove otherwise than by trial, that this or that word is not such, as it is guessed to be by others, according to their keys; [...] yet, if due trial being made, the key he proposes, shall be found so agreeable to the characters of the letter, as to enable one to understand them, and make a coherent sense of them, its suitableness to what it should decypher, is, without confutations, or extraneous positive proofs, sufficient to make it be accepted as the right key of that cypher.[51]

Due to its tentative nature the corpuscular hypothesis was even more viable than any rigid theory or ontological determination ever would have been. Its directing influence on future research was founded less on immediate results, than - as it was not uncommon in early modern science - on the basis of hope and promise. It was not until John Dalton's chemical atomism in the early nineteenth century, however, that this promise was to be fulfilled.

Institut für Geschichte der
Naturwissenschaften, Mathematik und Technik
Universität Hamburg

NOTES

1. In particular Kurd Lasswitz, *Geschichte der Atomistik vom Mittelalter bis Newton*, 2 vols. (Hamburg/Leipzig 1890).
2. Hélène Metzger, *Les doctrines chimiques en France du début du XVIIe à la fin du XVIIIe siècle* [1923] (Paris: Blanchard, 1969); G. B. Stones, "The Atomic View of Matter in the XVth, XVIth, and XVIIth Centuries," *Isis* 10 (1928), 445-465; Marie Boas, "The Establishment of the Mechanical Philosophy," *Osiris* 10 (1952), 412-541; Stephen Toulmin and June Goodfield, *The Architecture of Matter* [1962] (Chicago/London: University of Chicago Press, 1982), pp. 137-200; Ugo Baldini, "Il corpuscolarismo italiano del Seicento: Problemi di metodo e prospettive di ricerca," in: *Ricerche sull'atomismo del Seicento*, Pubblicazioni del 'Centro di Studi del Pensiero Filosofico del Cinquecento e del Seicento in Relazione ai Problemi della Scienza' I,9 (Firenze: Nuova Italia, 1977), pp. 1-76.
3. Lucr. IV, 116-122.
4. Lucr. II, 114-131.
5. Claudius Berigardus, *Circulus Pisanus de veteri et peripatetica philosophia* [1643] (Padova 1661), IV.8., pp. 421-422.

6. Daniel Sennertus, *Hypomnemata Physica* [1636], in: *Danielis Sennerti Opera Omnia*, vol. I (Lyon 1656), pp. 100-172, *Hypomnema III. De atomis et mistione* 1.1., p. 118.
7. Lucr. II, 391-397.
8. Iohannes Chrysostomus Magnenus, *Democritus reviviscens sive vita et philosophia Democriti* (Leyden 1648), p. 206-207; see also Alfred Stückelberger, *Antike Atomphysik: Texte zur antiken Atomlehre und ihrer Wiederaufnahme in der Neuzeit* (München: Heimeran, 1979), p. 282-285.
9. Walter Charleton, *Physiologia Epicuro-Gassendo-Charltoniana* (London 1654), II.4., p. 114.
10. Petrus Gassendi, *Syntagma philosophicum*, in: *Petri Gassendi Opera omnia in sex tomos divisa*, Vol. I (Lyon 1658), I.3.5., p. 265.
11. Alfred Stückelberger, "Lucretius reviviscens: Von der antiken zur neuzeitlichen Atomphysik," *Archiv für Kulturgeschichte* 54 (1972), 1-25, here on p. 19: "Zum erstenmal ist hier in der Geschichte der Atomphysik der Schritt vom qualitativen zum quantitativen Denken gemacht worden, mit einer übrigens durchaus brauchbaren Methode - später sind allgemeine Atomgewichte durch Gase errechnet worden -, die auch zu einer erstaunlich zutreffenden Größenordnung führt: er errechnet eine Anzahl von ca. $7{,}7 \cdot 10^{19}$ Atomen in einem Weihrauchkorn;" cf. idem, *Antike Atomphysik* (cit. n. 8), p. 60.
12. Jacobus Gaffarellus, *Curiositez inouyez, hoc est: Curiositates inauditae* [1629] (Hamburg/Amsterdam 1678), pp. 97-99; see also Lynn Thorndike, *History of Magic and Experimental Science*, vol. VII (New York: Columbia University Press, 1958), pp. 304-309.
13. Iosephus Quercetanus, *Ad veritatem Hermeticae medicinae ... responsio* (Paris 1604), pp. 292-294.
14. See *Correspondance du P. Marin Mersenne*, ed. by Marie Tannery and Cornelis de Waard, vol. I (Paris: Presses Universitaires, 1945), p. 326.
15. Petrus Gassendi, *Animadversiones in decimum librum Diogenis Laertii*, 2 vols. (Lyon 1649), I, p. 207; and verbatim also in idem, *Syntagma philosophicum* (cit. n. 10), I.3.6., p. 268-269.
16. Henry Power, *Experimental Philosophy ... in Avouchement and Illustration of the now famous Atomical Hypothesis* (London 1664), part I, sig. b2r, c2v-[c3r].
17. Ibid., observ. 34, p. 43.
18. Ibid., observ. 35, p. 44.
19. Ibid., p. 51.
20. Hans Kangro, "Erklärungswert und Schwierigkeiten der Atomhypothese und ihrer Anwendung auf chemische Probleme in der

ersten Hälfte des 17. Jahrhunderts," *Technikgeschichte* 35 (1968), 13–36, here p. 20.
21. Lucr. I, 298–299.
22. Lucr. I, 305–310.
23. Edward Grant, *Much Ado About Nothing: Theories of Space and Vacuum from the Middle Ages to the Scientific Revolution* (Cambridge/London/New York: Cambridge University Press, 1981).
24. Arist., Phys. IV.6. (213b21-22).
25. Francis Bacon, *Sylva Sylvarum*, I.34. [1627], in: *The Works of Francis Bacon*, ed. by James Spedding, Robert Leslie Ellis, and Douglas Denon Heath, Vol. II (London 1859), p. 354.
26. Gassendi, *Syntagma philosophicum* (cit. n. 10), I.2.3., p. 195; see also ibid., p. 196.
27. Ibid., p. 195.
28. Johannes Baptista Morinus, *De atomis et vacuo contra Petri Gassendi Philosophiam Epicuream* (Paris 1650), p. 19–20.
29. Lucr. I, 385–397.
30. Galileo Galilei, *Discorsi e dimostrazioni matematiche* [1638], in: *Le Opere di Galileo Galilei*, Edizione Nazionale, vol. VIII (Firenze: Barbera, 1965), p. 59.
31. Ibid., p. 62–63.
32. Arist., *Phys.* IV.9. (217a20-b19).
33. Gassendi, *Animadversiones* (cit. n. 15), I, pp. 424–444.
34. For his sponge-like concept of rarefaction see Renatus Descartes, *Principia philosophiae* [1644], II.6.-7., in: *Oeuvres de Descartes*, ed. by Charles Adam and Paul Tannery, vol. VIII/1 (Paris : Vrin, 1982), pp. 43-44. See also Boas, "Mechanical Philosophy" (cit. n. 2), pp. 442-460.
35. See W. Subow, "Zur Geschichte des Kampfes zwischen dem Atomismus und dem Aristotelismus im 17. Jahrhundert (Minima naturalia und Mixtio)," in: *Sowjetische Beiträge zur Geschichte der Naturwissenschaft*, ed. by Gerhard Harig (Berlin: Deutscher Verlag der Wissenschaften, 1960), pp. 161–191; Norma E. Emerton, *The Scientific Reinterpretation of Form*, Cornell History of Science Series (Ithaca/London: Cornell University Press, 1984), esp. pp. 76–105.
36. R. Hooykaas, "The Discrimination between 'natural' and 'artificial' Substances and the Development of Corpuscular Theory," *Archives Internationales d'Histoire des Sciences* 1 (1948), 640–651, here on p. 643; idem, "The Experimental Origin of Chemical Atomic and Molecular Theory before Boyle," *Chymia* 2 (1949), 65–80, here on p. 77.
37. Daniel Sennertus, *De consensu et dissensu Galenicorum et Peripateticorum cum Chymicis* [1619], in *Opera omnia* (cit. n. 6), p. 230; see also Tullio Gregory, "Studi sull'atomismo del

Seicento, II: David van Goorle e Daniel Sennert," *Giornale critico della Filosofia Italiana* 45 (1966), 44-63, here on pp. 53-55.

38. Angelus Sala, *Anatomia vitrioli* [1609], 3rd. ed. (Leyden 1617), pp. 96-97.
39. Angelus Sala, *Chrysologia seu examen auri chymicum* (Hamburg 1622), III.6.5., sig. K3r, F6v-7r.
40. Sennertus, *Hypomnema* III.1.1. (cit. n. 6), p. 118.
41. Ibid., p. 118-119.
42. E.g. Georgius Agricola, *De natura fossilium*, ed. by Georgius Fabricius (Wittenberg 1612), IX.2., p. 701.
43. Sennertus, *De consensu et dissensu* (cit. n. 38), II, p. 182.
44. Ibid., Preface to the reader.
45. Ibid., II, p. 182.
46. William Arthur Drumin, *The Corpuscular Philosophy of Robert Boyle: Its Establishment and Verification* , Ph.D. thesis, Columbia University, 1973, esp. pp. 79-83; see also Robert Boyle, "A Physico-Chymical Essay," in: *Certain Physiological Essays* [1661], in: *The Works of the Honourable Robert Boyle*, ed. by Thomas Birch, vol. I (London 1772), pp. 359-376, here on p. 375.
47. Thomas S. Kuhn, "Robert Boyle and Structural Chemistry in the Seventeenth Century," *Isis* 43 (1952), 12-36, here on p. 15; see also Boas, "Mechanical Philosophy" (cit. n. 2), pp. 494-499.
48. Peter Alexander, *Ideas, Qualities and Corpuscles: Locke and Boyle on the External World* (Cambridge/London/New York: Cambridge University Press, 1985).
49. Boas, "Mechanical Philosophy" (cit. n. 2), pp. 505-520; Arnold Thackray, *Atoms and Powers: An Essay on Newtonian Matter-Theory and the Development of Chemistry*, Harvard Monographs in the History of Science (Cambridge, Mass.: Harvard University Press, 1970).
50. Power, *Experimental Philosophy* (cit. n. 16), I, p. 58.
51. Robert Boyle, *About the Excellency and Grounds of the Mechanical Hypothesis* [1674, written ca. 1665], in: *The Works* (cit. n. 46), IV, pp. 67-78, here on p. 77.

ROBERT HALLEUX

THEORY AND EXPERIMENT IN THE EARLY WRITINGS OF JOHAN BAPTIST VAN HELMONT

Historians of science use to consider Johan Baptist Van Helmont (1579-1644) as a forerunner of modern experimental method, and indeed experiments are plentiful in his main work, the *Ortus medicinae* "Dawn of the medicine" of 1648.[1] However modern scholarship on him has raised serious doubts as to the originality and even the validity of his most famous procedures. Let me take a few examples.

- Experiment of the candle: in order to demonstrate that fire is not an element but a burning fume (*fuligo accensa*), Van Helmont set fire to a candle, extinguished it, and in the ascending smoke put the flame of another candle, which lighted the first.[2] This experiment has been borrowed from the XIth century alchemist Artephius[3], who had copied it from the Meteora of Aristotle.[4]
- The use of the thermoscope: in his controversy with De Heer about the composition of Spa water, Van Helmont intended to show that water is not transmuted into air, he used the thermoscope that is to say two glass balls connected by a tube.[5] He confessed himself that this device was first conceived as a perpetuum mobile.[6] In fact, it derived from a gadget built by the Flemish engineer Cornelius Drebbel[7] and is related to both the weather-glass of Fludd[8] and to the pelicane of the medieval alchemists.[9]
- The famous willow-tree experiment. Van Helmont attempted to show that water was the constitutive principle of living beings. He filled a bucket with mould-earth carefully sifted and weighed, and he planted a willow-tree which he watered during ten years. At the end of the ten years the weight of the earth had not significantly decreased and he might conclude that the growth of the tree was due to water only.[10] However, many scholars have shown that this first quantitative experiment in biochemistry has been copied from Nicholas of Cusa[11], and Christian Hunemörder has recently traced its origins back to a famous hermeticist romance of late Antiquity, I mean the *Recognitiones* of Pseudo-Clement of Alexandria.[12]

D. Batens and J. P. van Bendegem (eds.), Theory and Experiment, 93–101.

Similar doubts may be expressed about the use of the balance and the quantification of experiments. For instance, the comparison between the weight of a piece of wood before and after burning[13] is a well-known anecdote of the Greek polygraph Lucian of Samosata[14], the compression of air in a tube[15], the extinguishing of a candle under a glass bell (a cheese-cover)[16] came from the tradition of Greek and Renaissance mechanics (the Theaters of machines).

Other experiments, although easy, are clearly wrong, for instance a glass vessel is filled with water and hermetically sealed. It is weighed before and after freezing. Van Helmont observed the vessel was heavier before freezing.[17]

Finally several procedures are far more puzzling. "I have produced mice by spontaneous generation through fermenting of corn with dirty underwear in a sealed glass vessel".[18] Or "I have held in my hands a yellow-red powder with a grain of which I transmuted one thousand pounds of quicksilver into perfect gold and silver, and many people attended this experiment".[19]

Such ambiguities clearly rise the problem of the nature and the status of experiment. A comprehensive enquiry would involve a thorough survey of the Helmontian corpus and a threefold research.

a) The evidence of terminology, what is meant by *experientia, experimentum*.
b) Which status does he give theoretically to this kind of processes in his general theory of knowledge, that is to say in his methodological or epistemological texts.
c) Which role do play individual experiments in his scientific discourse, that is to say, in the building of his system.

This large scale enquiry far exceeds the scope of the present paper and will be the theme of my seminar next year. I shall restrict my purpose to a smaller group of texts written before the *Ortus*. I mean

- The *Eisagoge*, that is to say - *The introduction to the art of medicine restored by Paracelsus* a text written between 1599 and 1607, confiscated by the Holy Inquisition in 1634. Published by Broeckx in the XIXth.[20] I am now preparing a new critical edition under the auspices of the Koninklijke Akademie.
- *The treatise on the magnetic healing of the wounds* of 1609 published in 1621.[21] A short tract devoted to the weapon salve of Paracelsus.
- *The treatise on the fountains of Spa* of 1624 dealing with the analysis and medicinal properties of mineral waters.[22]
- The *de lithiasi* of 1643 over the stone in the bladder and the kidneys.[23]

1. THE EVIDENCE OF TERMINOLOGY

The necessity of clarifying first the meaning of words is not evident at first sight. However the French has only one world, *experience*, the English two, *experience* and *experiment*, the Latin a lot, *experientia, experimentum, probatio, periculum, examen*. According to the well-known philosophical vocabulary of Lalande *experience*,[24] in a general sense, means a sort of knowledge which is not innated but acquired through active of passive contact with reality. In a more technical sense, *experiment* is an observation which is created or provoked on the base of certain precise conditions. The result of this observation cannot be foreseen and is able to clarify the nature or the law of the so called phenomenon.

In Greek the term equivalent to experience is ἐμπειρία which has been defined by Aristotle in the *Metaphysics*.[25] Despite the opinion of the moderns, πεῖρα is not equivalent to experiment. A student of mine has analyzed all the occurrences of πεῖρα in ancient Greek and conclusively shown that it only means attempt, trial, and in a technical sense *testing* of substances, metals drugs and so on. So the first aphorism of Hippocrates does not mean "Experiment is doubtful" but only "art is long, life is short, trial of drugs on patients is dangerous".[26]

In Latin *expertum, experimentum* is equivalent to both ἐμπειρία and πεῖρα: it means both *experience* and *assaying*, *trial* but in the ancient and mediaeval medical tradition this word rises many ambiguities. For a lot of medical recipes and processes you can find *expertum est*. It means "it has been assayed, and it was successful".[27]

The term is especially used for technical or medical processes which are not rationally justified and for which no other evidence of validity can exist unless the success. I mean magical recipes. For instance the *experimenta Alberti Magni* are the famous secrets of Albert the Great[28]; the *experimenta magistry nicolai* are ugly remedies using snakes and toads[29], *the experimenta de coloribus* are illuminator's skills.[30]

So an *experimentum* is not a procedure which is planned in order to discover the rational order of things. It is a merely contingent fact, which succeeded once or which is said to have succeeded. This criterion is very loose, because there is no way of excluding coincidence or misinterpretation.

Moreover, most of the *experimenta* are guaranteed by a witness, who usually is not a person of the common. So a process experimented by a king or a pope is better and more powerful than a process tried by a peasant.

What now about Van Helmont? Those words occur from time to time but in the most general sense "statement of a good or bad result; medical knowledge acquired from facts and practice". For instance the academic physicians which he criticizes have a *longa aetatum experientia*[31]; *experientia medicorum et aegrorum*[32]; *aegris experientia teste*[33]; *experientia trivialis*.[34]

But three different words occur for the procedures: *experimentum, mechanica probatio "mechanical demonstration" quaerere per ignem* "searching through fire".

Experimentum keeps its mediaeval meaning. Van Helmont never used it for his own laboratory practice. Let me give you two examples: it is well known by experience that if you rub yourself with the fat of a seacalf, you won't be struck by thunderbolt[35]; it is also well known by the experience of the peasants that if you enemy defecates before your door, set a glowing iron into the excrements and your enemy will get scabies on his bottom.[36]

Mechanica probatio is used for a proof taken from the laboratory. It properly means "from the *mechanical* arts", opposed to the *liberal* ones.[37] Van Helmont is well aware of the long-lasting despise against this type of argument.[38] Let me remind you of Albert the Great's criticism of the alchemist ibn Juljul, *Mechanicus et non philosophus* who based his theory of metal genesis upon a technical comparison with glass-making.[39]

Last, *quaerere per ignem* qualifies the paracelsian methods of chemical fire analysis that is to say the spagyrical processes.[40] The Flemish is far more explicit : he considers himself as a fire craftsman, a *vuurkonstenaar*.[41]

2. THE THEORETICAL ACCOUNTS OF EXPERIMENTS

So we have three kinds of so-called "experiments": mediaeval *experimenta*, demonstration with gadgets, chemical analysis.

The epistemological status of *experimentum* and *analysis per ignem* versus theory and deductive reasoning is discussed at length in the *Eisagoge*. This text, confiscated by the Inquisition, is a comprehensive survey of *Paracelsian Theory*, beginning with a dream where the ghost of Paracelsus reveals him his secrets.

The first chapters are a history of knowledge.[42] According to Paracelsus, knowledge has to be based on the twofold light of God and nature. The earliest physicians, for instance Hippocrates, were inspired by both. After him, the birth of professional medicine involved struggle for power and wealth. This caste obscured the light of nature through demonstration *more geometrico*, by logic, sophisms, and building of *theoremata* (a loan-word coming from the technical vocabulary of the Greek empiricists).[43]

Galen and Aristotle were responsible for the new kind of medicine, reasoning on diseases instead of healing them. In fact, the only usefulness of medical theory is to support the claims of the medical establishment.

This fallacious building has been broken up by the irruption of new diseases, unknown of the ancients and reluctant to classical medicine. Scabies is reluctant to scabious hypotheses. Against these diseases, the *experimentum* of an old witch is better than the canon of Avicenna.[44]

According to Van Helmont this failure of theoretical medicine gave rise to the neo-empirical thought of Jean Fernel in his *de abditis rerum causis*.[45] For Van Helmont, this work is a mere collection of experimenta, that is to say in the mediaeval sense, magical processes the effects of which are ascribed to occult qualities. Van Helmont does not raise any doubt about the processes themselves, he only questions the lack of explanation.[46]

Last came Paracelsus, grand-grandson of Hermes. To this rough statement of facts, he added the resolution of bodies though art *(artificiosa resolutio)*. Chemical analysis enables the philosopher to separate the substances and to isolate the principles of things. For instance, the narcotic effects of poppy are no more explained by *vis dormitiva* but by a particular kind of *sulphur*.[47]

And he concluded with a reminiscence of the Danish paracelsist Peder Sörensen (Petrus Severinus) "Go ahead my sons, climb the mountains, gather flowers and minerals, build a furnace, keep awake in the laboratory. There is no other way to truth".[48]

The dramatic style of this call for a new experimental method may not overshadow its limits. Even in the context of the new learning the mediaeval *experimentum* keeps it validity and Van Helmont keeps admitting all sorts of facts and fancies without any criticism.

3. THE ROLE OF "EXPERIMENTAL" PROCEDURES IN THE SCIENTIFIC DISCOURSE

Let us switch now to the role played by individual procedures in the scientific discourse. I mean here mediaeval *experimenta*, mechanical proofs, chemical analysis.

Let us read the *De magnetica vulnerum curatione* of 1621. This treatise is full up with experiments and laboratory demonstrations. This treatise deals with the famous weapon-salve of Paracelsus. One had not to anoint the wounds; but the weapon which caused them.[49] The weapon salve controversy is important for the problem of action at a distance.[50] Van Helmont endeavoured to demonstrate that the effects of the salve are neither magical nor astrological but natural. Such a theory can only be substantiated by empirical evidence.

In fact, the pattern of reasoning is merely analogical. All sort of facts, popular beliefs and true experiments are advanced. For instance, the experiments of William Gilbert on the magnetic needle and the compass are at the same[51] demonstrative level as the corpses which bleed when the murderer comes.[52]

On the other hand the research with fire, *querere per ignem* provides the first draft of an experimental method. In the treatise on Spa Waters, Van Helmont proclaimed that *distillation* is the only way of enquiry because it put at before the eyes with certainly the inner structure of bodies.[53] For instance, distillation enabled him to identify in Spa waters only *aqua fontana*, fountain water and *vitriolum ferri (iron sulphate)*, in contrast to the long list of minerals found by the contemporaries.[54]

Last, the *de lithiasi*, treatise of stones shows the rise of a real experimental methodology. The problem is to know the coagulating agent involving stone formation in the bladder. Only one way, distillation "the mirror of intellect".[55] As suffering from stone himself, he distilled several times his own urine, varying the conditions of experiment; and isolated a salt, this salt was systematically tried in order to ascertain its dissolving or coagulating properties.[56] On the other hand, Van Helmont attempted to discover the same principle at work in similar processes of "coagulation" e.g. fossils, petrifications, glass, shells.[57] But the true interpretation of the result was only obtained after a prayer and ascribed to a gift of God.[58]

These few examples can at least suggest that the history of experiment in the Scientific Revolution is far less simple than it was thought. Lexicographical study and typological classification of the procedures seem to be the best way to avoid misunderstandings. In the early writings of Van Helmont, "ancient" and "modern" form of experiment can coexist: medico-magical *experimenta*, Renaissance gadgets and machines; assaying practice and, in the framework of distillation, the first trends of a method of enquiry, based upon organized and justified experiments.

Université de L'Etat à Liège

NOTES

1. J.B. Van Helmont, *Ortus medicinae, id est, initia physicae inaudita Progressus medicinae novus, in morborum ultionem, ad vitam longam*, Amsterdam, Louis Elzevir, 1648, reprint Brussels, Culture et Civilisation, 1966. A Flemish version of this text (which was in fact the original draft) was published later under the title *Dageraad, ofte nieuwe opkomst der geneeskonst, in verborgen grond-regulen der Nature*, Rotterdam, 1659. A preliminary list of "experiments" is found in H. De Waele, J.B. Van Helmont, Brussels, n.d. p. 52–57.
2. *Formarum ortus*, 29, *Ortus* p. 137.
3. Artephius, *Clauis sapientiae*, t. I, p. 508 B. Manget. See also Bruno, *De magia*, in *Opera*, t. III, p. 421.
4. Aristotle, *Meteorologica*, I, 4, 8, 342 a 3 sq.
5. *Aer*, 12–13, *Ortus*, p. 64–65 (quoting Clavius).
6. *Ortus*, p. 64 Dicit namque (de Heer) me voluisse edere motum perpetuum.
7. H. Michel, *Le mouvement perpétuel de Drebbel, Physis*, 13, (1971), p. 289–294.
8. A.G. Debus, *Key to two Worlds. Robert Fludd's Weather-glass*, dans *Annali dell' Istituto e Museo di Storia della Scienza di Firenze*, I (1982), p. 109–144.
9. Described by John of Rupescissa, *De consideratione quintae essentiae*, Basel, 1561.
10. *Complexionum atque mistionum elementalium figmentum*, 30, *Ortus*, p. 108–109. The same experiment is described with some discrepancies in *Dageraad*, p. 60. ("'t is bij proeve gesien").
11. R.C. Benedict, *The first experiment in plant physiology, Science*, 89 (1939) p. 411-412; H.E. Hoff, *Nicolaus of Cusa, van Helmont and Boyle: the first experiment of the Renaissance in quantitative biology and medicine, Journal of the History of Medicine*, 19 (1964), p. 99–117; H.M. Howe, *A Root of Helmont's Tree, Isis*, 56 (1965), p. 408–419; M.T. Walton, *Boyle and Newton on the transmutation of water and air from the root of Helmont's tree, Ambix*,27 (1980), p. 11–18.
12. C. Hünemörder, *Studien zur Wirkungsgeschichte biologischer Motive in den Pseudo-Klementinen, Medizinhistorisches Journal*, 13 (1978), p. 24–28.
13. *Complexionum atque mistionum elementalium figmentum*, 13, *Ortus*, p. 106, *Dageraad*, p. 166 (quoting Paracelsus).
14. Lucian, *Demonax*, 39.
15. *Supplementum de Spadanis fontibus*, II, 17, *Ortus*, p. 690; *Aer*, *Ortus*, p. 66; *Vacuum naturae*, 4, *Ortus*, p. 84.
16. *Vacuum naturae*, 7–9, *Ortus*, p. 84–85.

17. *Gas aquae*, 34-35, *Ortus*, p. 79; *Dageraad*, p. 217-218, See T.S. Patterson, *Van Helmont's ice and water experiments, Annals of Science*, I (1936), p. 462-467.
18. *Imago fermenti impraegnat massam semine*, 9, *Ortus*, p. 113.
19. *Arbor vitae, Ortus*, p. 793.
20. J.B. Van Helmont, *Eisagoge in artem medicam a Paracelso restitutam*, ed. C. Broeckx, *Le premier ouvrage de J.B. Van Helmont, Annales de l'Académie d'Archéologie de Belgique*, 10 (1853), p. 327-392; II (1854), p. 119-191.
21. *De magnetica vulnerum curatione*, Paris, Le Roy, 1621, reprinted (without any change) in *Ortus*, p. 746-780.
22. *Supplementum de spadanis fontibus*, Liège, Streel, 1624, reprinted (without any change) in *Ortus*, p. 685-704.
23. *De lithiasi*, in *Opuscula medica inaudita*, Amsterdam, Louis Elzevir, 1648.
24. M. Lalande, *Vocabulaire technique et critique de la philosophie* , 9e éd., Paris, 1962, p. 321-323, s.v. expérience.
25. Aristotle, *Metaphysics* , A., 1, 980 a 27-29, 980 b 26-90.
26. Hippocrates, *Aphorisms*, I, 1, t. IV, p. 458 Littré.
27. For the meaning of *exper* - in philosophical texts, see Pierre Michaud-Quantin, *Etudes sur le vocabulaire philosophique du moyen âge*, Roma, 1970, p. 219-221, for the medical texts C. Opsomer, *Prolégomènes à une étude des recettes médicales latines, Mémoires du Centre Jean Palerne,* III (1982), p. 94-96.
28. L.Thordike, *Further considerations on the Experimenta, Speculum astronomiae and De secretis mulierum ascribed to Albertus Magnus, Speculum,* 30 (1955), p. 413-443.
29. J.W.S. Johnsson, *Les Experimenta magistri Nicolai, Bulletin de la Société Française d'Histoire de la médecine*, 10 (1911), p. 269-290.
30. Ed. M.E. Merrifield, *Original treatises on the art of painting*, London, 1849, t. I, p. 49.
31. *Eisagoge*, p. 351.
32. *Eisagoge*, p. 347.
33. *Eisagoge,* p. 132.
34. *Eisagoge*, p. 171.
35. *De magnetica,* 45, p. 757.
35. *De magnetica,* ibid.
36. *De magnetica,* 21, p. 752.
37. *De spadanis,* II, 17, p. 690; VII, 16, title p. 699; *De lithiasi* , III, 42, p. 32; IV, 3, title, p. 32; *Terra* 3, title, *Ortus*, p. 54; *Vacuum naturae*, 4, *Ortus,* p. 84.
38. F. Allesio, *La filosofia e le "artes mechanicae" nel sec. XII, Studi Medievali,* 6, (1965), p. 91-154.

39. Albertus Magnus, *De mineralibus*, III, 1, 4, p. 163 Wyckoff.
40. *De Spadanis*, VII, 22, p. 701; *De lithiasi,* III, 1, p. 20-21 inter orare ergo et pulsare supponitur medium in naturalibus quaerendi per ignem; III, 30, p. 26 per ignem edocto.
41. *Dageraad*, p. 70 cf. p. 210. The corresponding Latin word is *pyrotechnia, Ortus*, p. 461, 462, 463 etc.
42. *Eisagoge*, p. 346-347.
43. See e.g. *Eisagoge*, p. 357, *De lithiasi*, III, 30, p. 29.
44. *Eisagoge*, p. 351.
45. Jean Fernel, *De abditis rerum causis Libri duo*, Paris, 1548. Fernel is quoted in *Eisagoge*, p. 359.
46. *Eisagoge,* p. 348 and 359.
47. *Eisagoge*, p. 347.
48. *Eisagoge*, p. 169, reminiscence of Petrus Severinus, *Idea medicinae philosophicae,* Basel, 1571. Other references, *Eisagoge,* p. 355; *De spadanis*, VII, 55, p. 704.
49. The origin of the Weapon-salve controversy was the tract of Rodolf Goclenius, *Oratio qua defenditur vulnus non applicato etiam remedio citra ullum dolorem curari naturaliter posse,* Marburg, 1608.
50. See my *Helmontiana, Academiae Analecta,* 45 (1983), p. 35-63.
51. William Gilbert, *De magnete*, London, 1600. Quoted in *De magnetica*, 19, p. 751; 39, p. 756; 65, p. 761 B. see A.G. Debus, *Robert Fludd and the Use of Gilbert's De magnete in the weapon-salve controversy, Journal of the History of Medicine*, 19 (1964), p. 389-417.
52. *De magnetica*, 113, p. 770 cf. 62, p. 760.
53. *De Spadanis*, VII, 22, p. 701. Quapropter Analysis per ignem adeunda est, quae sane ut reclusionem corporum, ita eorundem compagem certam, ante oculos proponit, certioreque spondet studio, quam quae de materia prima, priuatione, fortuna, casu, infinito, et vacuo, insomnia, jejuno adhuc ore, adolescentum verterit doctrina.
54. *De spadanis*, IV, 1-3, p. 693-694.
55. *De lithiasi*, III, 1, p. 21.
56. *De lithiasi*, III, p. 21-26 - This coaqulating principle was called duelech. cf. *Dageraad*, p. 238 sq.; L. Elaut, *Op zoek naar een etymologie voor Paracelsus' en J.B. Van Helmont Duelech, Fachprosa-Studien*, Berlin, 1982, p. 545-551.
57. *De lithiasi*, II, 13, p. 19; III, 32, p. 29.
58. *De lithiasi*, II, 15, p. 20; III, 32, p. 29-30.

NILS-ERIC SAHLIN

THE SIGNIFICANCE OF EMPIRICAL EVIDENCE FOR DEVELOPMENTS IN THE FOUNDATIONS OF DECISION THEORY*

1

This paper concerns the question of the significance of empirical evidence for developments in the foundations of decision theory. I will argue that a particular type of empirical findings is of crucial importance for how the theory develops. My position will thus be in sharp contrast to what I take to be the general view of the matter, i.e. that empirical findings have been and should be without significance for developments in the foundations of decision theory. In order to be able to outline more fully my aims and ideas a survey of the traditional theory of decision making, the Bayesian theory, is called for.

According to the Bayesian theory, a decision is a choice of one action alternative from a set of available alternatives. These possible actions are thought to lead to various outcomes depending on how the world develops. It is commonly assumed that the decision maker can numerically evaluate the utility of the outcomes in a decision problem and that his or her beliefs about the developments of the world, about the possible states of the world, can be represented by a unique probability measure. The fundamental decision rule of the Bayesian theory is the principle of maximizing expected utility: In a given decision problem, choose the alternative with the maximal expected utility.

The Bayesian theory has traditionally been presented as a system of axioms. These axioms are used to give the theory a solid foundation and from them various representation theorems can be proved; e.g. that there exists a subjective utility measure, $P(.)$, defined over the possible states of the world which represents the decision maker's degrees of belief; that there exists a utility measure, $u(.)$, defined over the possible outcomes, determined up to a positive linear transformation; and, finally, that if we act in accordance with these axioms, then we will follow the principle of maximizing expected utility.

The best known axiomatizations of the Bayesian theory are those of Ramsey (1931), Savage (1954), and Jeffrey (1965). It is frequently forgotten that these three theories are in some respects

D. Batens and J. P. van Bendegem (eds.), Theory and Experiment, 103–121.

substantially different. We should, if we are to be correct, talk about Bayesian theories rather than about *the* Bayesian theory. However, by a Bayesian theory we often just mean a theory comprehending such ideas as subjective probability, a utility measure, and a belief in the appropriateness of, in one way or another, maximizing expected utility.

A system of axioms, as those used, for example, by the Bayesian theories, can be divided into two categories; structure axioms or ontological axioms, on the one hand, and rationality or behavioural axioms, on the other (see Suppes, 1956).

The structure and ontological axioms are mostly of purely mathematical or philosophical interest. They tell us, for example, what we can expect to exist, and they provide us with the necessary mathematical tools to prove the desired representation theorems. A good example of an axiom of this kind is Ramsey's first axiom which states that there "is an ethically neutral proposition *p* believed to degree 1/2" (p. 178). Another archetype of this sort of axioms is the archimedean axiom.

The rationality axioms are axioms that a "rational" person is supposed to satisfy when making decisions. A typical example of a rationality axiom is the transitivity principle. Stated in terms of outcomes this principle says that for all outcomes *a*, *b* and *c*; if *a* is preferred to *b*, *b* is preferred to *c*, then *a* should be preferred to *c*. The best-known argument in favour of this principle or axiom is the so-called "money pump" argument. Assume that you violate the transitivity principle: you prefer *a* to *b*, *b* to *c*, but, you prefer *c* to *a*. Thus, given *b*, we would expect that you are willing to pay a small sum of money (as small as you like) to get *a* instead. Similarly, you would be willing to pay an amount of money to get *c* instead of *a*, since you prefer *c* to *a*, but also to pay money to get *b* back, since *b* is preferred to *c*. Thus, you end up with what you had at the beginning, but with less money. If you do not correct your intransitive preferences, you can be used as a money-pump until you run dry. This and similar types of arguments are put forth in favour of most of the rationality axioms.

From an empirical point of view it is the rationality axioms that are of interest. These are the axioms that can be empirically tested (the distinction between the two kinds of axioms is somewhat slippery, structural and ontological axioms can have behavioural significance, e.g. if we want take cognitive limitations into account) and psychologists have thoroughly scrutinized them. An overwhelming majority of these experiments indicate that people are far from satisfying even the most common and acceptable rationality axioms (see e.g. Kahneman, Slovic & Tversky, 1982). This fact has led to a debate on whether people are rational or not. I would argue that these experiments do not tell us much about human rationality. What they do tell us is that we generally do not act in accordance with a number of, what we believe to be, reasonable principles of sound behaviour. The question of rationality can, I take it, only be answered within the framework of much more general psychological theories. I will therefore, as far as possible, try to avoid talking about rational decisions and instead use the word "sound decisions."

Another question, and the central question of the present paper, concerns the significance of this type of empirical findings for the foundations of decision theory. Theories of decision, choice or utility, are generally thought, because of their normative status, to be unaffected by empirical evidence. "People can do what they want, they are irrational, our theory is only for those of us who are truly rational," we seem to be arguing. But, for whom then do we invent these metaphysical theories - immune, as they are, to empirical findings? I will not further discuss this position, untenable as it is. My own view of the matter is that empirical evidence is of great significance for developments of normative theories, or better, for the developments of prescriptive theories; theories telling us what is sound behaviour in a given situation. A prescriptive theory, with its underlying intent to provide us with a basis for sound behaviour, should be tested against our intuitive understanding of what a sound decision is but also against what we actually do. To lift a line from Ramsey one could say that a prescriptive theory should be as "a map of neighboring space by which we steer" (1931, p. 238). If this map misguides us because it is not properly drawn or because we can't read it, then we had better redraw it.

The history of decision theory clearly shows that there have been empirical findings which have had a great impact on developments in the foundations of the theory. However, most of the empirical results published in this area have had no significance at all. In the following two sections, I will give a couple of examples of empirical findings which have been without significance and put these in contrast to results that have suggested sweeping changes of the theory. This case study reveals that the latter type of findings, the very few results that occupy a place apart, seems to satisfy a number of criteria which are not satisfied by the majority of empirical results on decision making. These criteria will be specified and I will argue that only those findings that actually satisfy the criteria should have any influence on developments in the foundations of decision theory (or, more accurately, on the foundations of any prescriptive theory).

2

This section is devoted to two experiments on human decision making that have had little or no significance for developments in the foundations of decision theory. I have not selected two uninteresting or scamped pieces of work. On the contrary, I have tried to find two first class experiments, with a carefully planed design, important results and a well reasoned analysis of the data obtained. Furthermore the experiments have to some extent influenced subsequent research in the area. They are archetypes of a set of experimental findings that should make us wonder: Why is it that we neglect these important results when we aim at developing our theories?

The transitivity principle is one example of a rationality axiom. It is a compelling principle of sound behaviour, substantiated by strong theoretical arguments, especially through the "money-pump" argument. In short, it is the type of principle of sound choice that most of us want to, and believe that we do, satisfy. However, experimental findings indicate that we do not behave in accordance with this principle. Our preferences are often, at least those involving some degree of complexity, intransitive. It seems as if we are following a strategy of intransitivity rather than a principle of transitivity, a fact disclosed, for example, by the experimental results discussed in Tversky (1969).

In one of his experiments Tversky investigated subjects' preferences between simple gambles. The gambles employed were of the form win-or-nothing (depending on a chance event one either receives a few dollars or nothing). In order to avoid expected values being computed by the subjects, the probability of the chance events was visualized as black sectors on a disc. The five gambles used in the experiment was constructed so that evaluation of payoff differences was easier than that of probability differences. The probability of winning and the prize received for the five gambles (*a* to *e*) were, 7/24 to 11/24 (in 1/24 steps) and $5 to $4 (in -$0.25 steps), respectively. Note that the design implies that expected value is negatively correlated with payoff.

Tversky conjectured that some of the subjects would ignore small probability differences between gambles and that, provided the probability or value difference between two gambles was small enough (a step a part), they would choose on the basis of payoff. However, for gambles with distinct probability or value differences it was hypothesized that the subject would choose according to expected values; because of the negative correlation between expected value and payoff. To give an example let us take three of the above gambles: *a*=(probability 7/24 to win $5), *b*=(probability 8/24 to win $4.75) and *c*=(probability 9/24 to win $4.50). One might expect that *a* is preferred to *b*, because of *a*'s better payoff, and that, for the same reason, *b* is preferred to *c*. But *c* may be preferred to *a* all the same, because of its higher expected value and despite its lower payoff. This type of preference pattern will, in general, induce a violation of the transitivity principle.

Ringing the changes on this theme, Tversky found that under specified experimental conditions, systematic, consistent and predictable patterns of intransitive choices could be obtained. One type of choice theory, which under some conditions violates the transitivity condition, is the so-called additive difference model. If one could find out the precise form of the difference functions of this model, it could well serve as an inductive generalization of the observed behaviour, thus allowing for a description, in a precise mathematical form, of the intransitive behaviour of some people.

Tversky emphasizes that he had no intention of addressing the question of whether human preferences are, in general, transitive. However, in the light of other empirical investigations, directly or indirectly concerned with this question, it seems safe to argue that, to some extent, people have intransitive preferences.

We thus have rather strong empirical evidence against one of the basic principles of sound behaviour or choice. But this fact has not led us to look for new axioms or theories of decision making. The experimental results on transitivity, regardless of their descriptive force, have had little or no significance for developments in the foundations of decision theory. The mathematical models these findings resulted in are not considered as serious alternatives to the traditional theories. In section 4 I will return to this example and try to explain why I believe that this is as it should be. However, before I turn to a number of empirical results which have had a notable influence on the theoretical developments in the foundations of decision theory, I want to conclude this section with a different example of a type of experimental results of purely psychological interest.

Lichtenstein & Slovic (1971) found a reversal of preference between bids and choices in gambling decisions. In their experiment, gambles with excellent odds (e.g. 0.99 probability of winning $4 and 0.01 probability of loosing $1: expected value = $3.95) were combined with gambles with large winning payoffs (e.g. 0.33 probability of winning $16 and 0.67 probability of losing $2: expected value = $3.94), and the subjects were asked to pick, for each such pair, the gamble they would prefer to play. Lichtenstein

& Slovic compared this information with the subjects' willingness to sell individual gambles, i.e. the subject was told that he or she owned a ticket to play a specific gamble and was then asked to state the minimum selling price for this ticket. The results obtained revealed that people frequently chose one gamble from a pair of gambles, but, at the same time, bid more for the gamble they did not choose. The two gambles above may serve as an example. A gamble with a 0.99 probability of winning seems to be preferable to a gamble with only a 0.33 probability of winning. However, if the two gambles were for sale, it seems reasonable to charge more for a gamble with a large winning payoff, $16, than for a gamble with a payoff four times lower. These results indicate that for themselves subjects prefer security but to others want to sell opportunities. It would take some ingenuity to revise the traditional theories in such a way that they could account for this type of behaviour. This would also be a pointless enterprise since no one really believes that these results should have any effect on our theories. But why do we so strongly believe this?

3

Another rationality axiom that has come under heavy fire is Savage's so-called sure-thing principle. This axiom asserts that if two alternatives have a common outcome under a particular state of nature, then the ordering of the alternatives should be independent of that outcome. If I can take out an insurance policy from either company A or B and I observe that the two companies have the same terms for a particular type of accident, then this fact should be of no importance for the decision I am about to make.

The sure-thing principle has commonly been recognized as a compelling principle of rational behaviour. It appears as an axiom, sometimes formulated somewhat differently, in a number of theories of choice. To give but one example: Assuming that we accept objective probabilities (some people do), it can be shown that together with some other assumptions the sure-thing principle logically implies one of the best-known principles of game or utility theory – the so-called independence axiom.

Savage says that he knows of "no other extralogical principle governing decisions that finds such ready acceptance" (1972, p. 21). But, the sure-thing principle is far from universally accepted. The principle, and the Bayesianism it is part of, have been criticized through a conceptual analysis of its underlying assumptions and through more descriptive psychological consideration. Below I will outline this influential criticism and try to indicate how it has led us to look for new foundations of sound behaviour.

Ellsberg's paradox. One of the best arguments against the sure-thing principle, and the Bayesian assumption that a decision maker's state of knowledge can be represented by a unique probability measure, is due to Daniel Ellsberg (1961).

> Ellsberg asks us to consider the following decision problem. Imagine an urn known to contain 30 red balls and 60 black or yellow balls, the latter colours in unknown proportion. One ball is to be drawn, at random, from this urn. The first choice situation consists of two alternatives a_1 and a_2. If you choose a_1 you receive \$100 if a red ball is drawn and nothing if a black or yellow ball is drawn. If you choose a_2 you receive \$100 if a black ball is drawn, otherwise nothing. Under the same general design, a second choice situation consists of the two alternatives a_3 and a_4. If you choose a_3, you receive \$100 if a red or yellow ball is drawn, otherwise nothing. If you choose a_4, you receive \$100 if a black or yellow ball is drawn, otherwise nothing. These two decision situations can be summarized by the following decision matrix:

	30	90	
	Red	Black	Yellow
a_1	\$100	\$0	\$0
a_2	\$0	\$100	\$0
a_3	\$100	\$0	\$100
a_4	\$0	\$100	\$100

Empirical investigations indicate that the most frequent pattern of response to these two decision situations is that a_1 is preferred to a_2 and a_4 is preferred to a_3. It is easily seen that this preference pattern violates the sure-thing principle.

These preferences also seems to be of the robust kind. Ellsberg writes: "The important finding is that, after rethinking all their "offending" decisions in the light of the axioms, a number of people who are not only sophisticated but reasonable decide that they wish to persist in their choices" (1961, p. 656). Being told that we have expressed a pair of "irrational" preferences, we have violated an axiom taken to be one of the fundamental principles of "rational" behaviour; most of us refuse to change our preferences.

We simply do not find the arguments in favour of the principle compelling enough.

Ellsberg's findings are supported by the experimental results of Slovic & Tversky (1974). Slovic and Tversky found that, if people were given lucid and persuasive arguments both in favour and against the sure-thing principle, there was an overwhelming preference for violating the principle. Furthermore, this study shows that it is far more likely that one who is acting in accordance with the axiom will change his or her mind and violate it, if given a second choice, than one who is violating the principle will come to accept it.

I take it that Ellsberg's problem provides us with strong empirical evidence against the sure-thing principle. One could, of course, argue that the only thing this type of experiment shows is that the axiom was not properly understood by the subjects and that what is needed is not a new theory of decision but rather education. I believe that it would be unfortunate if by referring to "education" and "understanding" we neglected the significance of these psychological findings for the foundations of decision theory. Firstly, even people who fully understand the content of the axiom do not accept it and some of them, as Ellsberg remarks, violate it with gusto. Secondly, there are a number of distinct but related empirical findings that pinpoint exactly the same cluster of problems connected with the traditional Bayesian theory.

What we should do is to see if our axioms of rationality and their theoretical consequences, in the light of these negative experimental findings, are still acceptable; can they, for example, derive support from more fundamental theoretical considerations?

The rationale for the preferences revealed in the Ellsberg problem seems to be that there is a difference between the quality of knowledge we have about the states. We know that the proportion of red balls in the urn is one third, while we are completely uncertain about the proportion of black (and yellow) balls (it can be anything between zero and two thirds).

The difference in information about the states is reflected in the preferences in such a way that the alternative for which the exact probability of winning can be determined is preferred to the alternative where the probability of winning is, in Ellsberg's words, "ambiguous." The ambiguity or unreliability, resulting from the amount and quality of information which the decision maker has about the possible states and the possible outcomes, is an important factor when making a decision, however, a factor which cannot be handled by the traditional Bayesian theory. This is due to the fact that a central theorem of the Bayesian theory says that the decision maker's knowledge can be represented by a unique probability measure. A simple theoretical argument emphasizes the conceptual drawbacks resulting from this fundamental tenet of how a decision maker's degree of belief is best represented.

Imagine you own a ticket to play the following gamble: If there is a transit strike in Verona, Italy, during the next month, you will receive $100; if not, you get nothing. For the sake of argument, assume that, after considering it carefully, you believe that the probability that there will be a transit strike in Verona next month is 0.30. Thus, provided that dollars and utilities are exchangeable, the (subjective) expected utility of this gamble is $30. In your opinion 30 dollars would be a fair price for the ticket.

Now assume that I offer to trade this ticket of yours for a ticket to play the following gamble: If you draw a black ball from an urn containing 30 black and 70 white balls, you get $100; if the drawing results in a white ball, you get nothing. With a 0.30 probability of success, this gamble has the same expected utility as the previous one. However, although the two gambles (for you) have the same expected utility, I am confident that you will trade the first gamble for the second. This preference conflicts with the recommendations of the Bayesian decision theory.

Why do you prefer the second gamble? Is it because you have assigned too high a value to the probability that there will be a transit strike in Verona? Let us assume it is not; 0.30 is, under the circumstances, the best assessment you can make. The crux of the matter is that you feel that you know more about the urn than about Italian wages, working conditions and other important factors that may provoke a transit strike in Verona.

This example is important. It shows that it is possible to find decision situations which are identical in all respects relevant to the strict Bayesian, but which nevertheless motivate different decisions. There is an important difference in degree between our knowledge concerning drawings from urns and transit strikes. As Kyburg has pointed out already (1961), one problem about the strict Bayesian view of probability is that it cannot account for this obvious difference in ignorance which prevails in the two gambles. Savage was certainly clear about the problem.

> To approach the matter in a somewhat different way, there seem to be some probability relations about which we feel relatively 'sure' as compared with others. When our opinions, as reflected in real or envisaged action, are inconsistent, we sacrifice the unsure opinions to the sure ones.

> The notion of 'sure' and 'unsure' introduced here is vague, and my complaint is precisely that neither the theory of personal probability, as it is developed in this book, nor any other device known to me renders the notion less vague. (1972, pp. 57-58)

Savage also mentions another, but related, theoretical difficulty with the Bayesian approach to probability.

> A second difficulty, perhaps closely associated with the first one, stems from the vagueness associated with judgments of the magnitude of personal probability. The postulates of personal probability imply that I can determine, to any degree of accuracy whatsoever, the probability (for me) that the next president will be a Democrat. (1972, p. 59)

What Savage alludes to is the fact that our beliefs in the statement "the next president (of the United States) will be a Democrat" can be fixed to any degree we like. This is a central theorem of the theory. But, isn't this asking too much? Precise probability assessments seem fine for well-determined situations, like drawings from such inventions as urns. However, are there any real-life events for which you can assess the probability with the same accuracy? What is, for example, the probability of rain in Ghent tomorrow?

My guess is that you will answer questions like this one by a one place or two place decimal number. I can probably squeeze a third and a fourth decimal number out of you, but you will emphasize that these assessments are very unreliable or ambiguous. A fifth and a sixth decimal assessment is out of the question.

I believe that this discussion shows that the Bayesian tenet that the decision maker's state of belief concerning which state obtains can be represented by a unique probability measure, and to any degree of precision, is too strong and ought to be relaxed. Thus, theoretical arguments together with, for the Bayesian theory, negative experimental findings, should have us look for new prescriptive theories of sound decision making.

One obvious way to handle the above difficulties is to deny that the decision maker's beliefs about the states of nature can be represented by a unique probability measure and instead argue that these beliefs are best represented by a set of such measures (intervals are, for theoretical reasons, in general not sufficient). This set is best conceived of as consisting of epistemically possible and reliable probability measures. That is as a basis for action the decision maker want to use measures that do not contradict his knowledge in the given situation, but also measures that are backed up by more information than other measures (for details see Gärdenfors & Sahlin, 1982, 1983, 1987, and Sahlin, 1983).

But decision theories based on probability assessments in terms of sets of probability measures cannot simply employ the traditional decision rule of maximizing expected utility. The reason for this is that for each measure in the set employed as basis for action we can calculate an expected value and thus each action alternative open to the decision maker will be associated with a set of such values. This problem can, however, be solved by introducing more complex decision rules. Different such rules has been suggested by, for example, Kyburg (1983), Levi (1974, 1980) and Gärdenfors & Sahlin (1982, 1983). One would expect that, employing these rules in a specific decision problem, they would more or less guide us towards the same decision. But, as is pointed out in Sahlin (1985) a simple example shows this is not the case. Thus, empirical findings and theoretical criticism have led us into the situation that, from having one answer to the question how to make a sound decision, we now have three competing answers.

Allais' paradox. My second example of a type of experimental findings that has had a great impact on the developments of decision theory may, at a first glance, look very similar to the findings of the Ellsberg type. However, I will argue that there is a difference in degree between how important these two types of findings are.

Maurice Allais suggested the following decision problem as an example of a situation in which one violates the sure-thing principle.

> Imagine a lottery with a hundred numbered tickets. One ticket is to be drawn at random. The first choice situation consists of two alternatives a_1 and a_2. If you choose a_1 you receive \$500,000 whatever ticket is drawn. If you choose a_2 you receive \$2,500,000 if one of the tickets numbered from 2 through 11 is drawn, and \$500,000 if one of the tickets numbered from 12 through 100 is drawn, otherwise you get nothing. Under the same circumstances, a second choice situation consists of the alternatives a_3 and a_4. If you choose a_3 you get \$500,000 if one of the tickets numbered from 1 through 11 is drawn, otherwise nothing. If you choose a_4 you get \$2,500,000 if one of the tickets numbered from 2 to 11 is drawn, otherwise nothing. These two decision situations are summarized in the following decision matrix.

	1	2–11	12–100
a_1	\$500,000	\$500,000	\$500,000
a_2	\$0	\$2,500,000	\$500,000
a_3	\$500,000	\$500,000	\$0
a_4	\$0	\$2,500,000	\$0

In this decision situation, many people prefer a_1 to a_2 and a_3 to a_4, and this preference pattern clearly violates the sure thing principle. We note, for example, that this pair of preferences implies a pair of incompatible utility inequalities; $0.11 \times u(\$500,000) > 0.10 \times u(\$2,500,000)$, by the first preference, but, at the same time, the reverse inequality holds by the second preference (assuming that $u(\$0)=0$).

But, this inconsistency is dissolved if one assumes that we employ distinct utility functions in the two decision situations. In the first situation we know that we, with certainty, can receive a fortune. It thus seems reasonable to evaluate the \$2,500,000 prize in the light of our possible prosperity, rather than with respect to our present state of poverty. The possible gift of \$500,000 changes our view of life, it changes our levels of aspirations. In the second situation there is no choice which will result in a fortune with certainty; thus we have to evaluate the possible prizes with respect to our present wealth. Taking the decision maker's level of aspirations (target levels or reference levels) into account, provides us with a ready explanation of Allais' problem.

Savage, however, thought that this type of "error" of preferences would be corrected if it was pointed out to the decision maker. When we realize that, if a ticket numbered from 12 through 100 is drawn, we get the same outcome regardless of what action is chosen, the two decision situations are symmetrical for tickets numbered from 1 through 11, then we will correct our preferences in such a way that they come to satisfy the sure-thing principle. But this is, unfortunately, not true, as is clearly shown by the experiments carried out by Slovic & Tversky (1974).

Besides Allais' paradox there is a number of other more or less well known violations of the sure-thing principle that have to do with the assessment and aggregation of utilities rather those with that of probabilities (see, for example, Kahneman & Tversky, 1979, and Machina 1982, 1983). Furthermore, the sure-thing principle, or its equivalent within the realm of utility theory, the independence axiom, appears to have less normative validity than we would like it to have. Analysis of these principles carried out by McClennen (1983), who investigates four different lines of argument that have been used to defend these cornerstones of the theories of rational choice, clearly shows that these arguments do not hold water.

Thus, analogous to the previous discussion of subjective probabilities, experimental results supported by conceptual analysis suggest that we ought to look seriously for new foundations of utility theory. One such alternative to the traditional theory has been suggested by Mark Machina (see, for example, 1982, 1983). I will not discuss this theory in any detail. But Machina, who finds the theoretical and empirical arguments against the independence axiom convincing, drops the independence assumption by changing the functional form of the preference functional. This step together with an assumption, which he calls *Hypothesis II* (stated in terms of Arrow's and Pratt's risk measure the hypothesis says that a shift from one probability measure to one stochastically dominating it yields a more risk averse utility function), enables him to give a systematic and unified analysis of people's seemingly unsystematic violations of the independence axiom.

4

The task in front of us is to explain why it is reasonable that experimental findings of the Allais and Ellsberg types have had a much greater influence on our search for new foundations of decision theory than experimental findings like the ones on transitivity or reversal of preference.

The first thing one should note is that empirical findings which are not substantiated by theoretical arguments, or theoretical arguments that lack empirical support, are unlikely to overthrow any theory of the kind we are discussing here. If people had accepted the sure-thing principle and chosen as Savage wanted them to do in Allais and Ellsberg problems, we would have been most happy with our theory and kept it as it is. A Kyburg or a McClennen would probably have had some difficulties making their voices heard. I believe that this is shown by the fact that although Kyburg's criticism of the Bayesian theory was published already in 1961, it is not until recently, and thousand of pages of experimental data later, that we take it seriously. Similarly, the experimental findings on their own would have had little effect on how we judged the theory. Thus it is clear that it is the concurrence of conceptual criticism and experimental findings that has been of significance and which is necessary.

But what type of empirical findings or conceptual criticism is needed? I believe the present case study reveals two main criteria and a number of subcriteria that are tacitly employed. We want both the empirical evidence as well as the theoretical criticism to be robust. Let me in more detail go through, in the light of the above case study, what is meant by this.

I will start with what I call *robustness of empirical arguments*, or shorter, empirical robustness. Only those empirical findings that are empirically robust should have (and, actually, have had) any significance for the development of a theory.

There are four aspects of empirical robustness that should be emphasized in the present context: the criterion of understandability, of cognitive stability, of persistence and, finally, of lack of alternative explanations. I will argue that the empirical results of the Allais and Ellsberg type satisfy these criteria, but that the findings on transitivity and reversal of preferences do not.

Firstly, the subjects should have a clear understanding of the task they are performing. They should have full comprehension of the axioms or of any theoretical argument presented to them. The experiments on the sure-thing principle, and with it related experiments, have been of this kind. The experiments of Ellsberg, Slovic & Tversky and others, show that this is the case. With respect to this criterion there is no difference between the experiments discussed in the previous sections. Also the experiments on transitivity and reversal of preferences satisfy this criterion.

Secondly, we have the criterion of cognitive stability. We want, for example, to make sure that our results are not simply due to an inability to handle and process vast amounts of information. It is not our cognitive limitations that should be in focus. An aspect of this criterion is that, even if the subjects understand the task they are performing, we must avoid their not simply trying to have it over and done with, i.e. we want to avoid the results being due to carelessness. I do not think that this problem is dodged simply by not using professional subjects, i.e. (psychology) students, but rather is a problem of experimental design. I take it that the empirical findings discussed here satisfy this criterion of cognitive stability or robustness.

Thirdly, the subjects should be persistent in their behaviour. They should not change their minds about what they view as the sound decision even if they are given strong arguments against their behaviour. The experiments carried out by Slovic & Tversky on Allais' problem and on Ellsberg's problem show that people are persistent in their violation of the sure-thing principle. They simply do not accept the arguments in favour of this principle. They strongly believe that the theory asks them to make the wrong decision. People do not, however, exhibit the same persistence when it comes to maintaining intransitive preferences. Most of us tend to believe that we are and should be transitive, despite the fact that, generally, we are not. We accept the transitivity principle, the arguments in its favour, and we believe that a violation of it is a violation of one of the fundamental principles of sound reasoning. This means that when our intransitive preferences are brought to light, we try to work towards transitive preferences. Thus, the findings in Allais and Ellsberg situations satisfy this criterion of persistence, but the results on transitivity do not.

Finally, we have the important criterion of lack of alternative explanations (within the theory) of the empirical data. By this I mean that, simply a minor shift of perspective (if you are hypersensitive, call it an ad hoc alteration of the theory) should not make it possible to explain (away) the empirical results. The results should be of a kind that demand a total reformulation of fundamental principles of the theory. The case study above clearly indicates that this is true for the empirical results on ambiguity or unreliability, provided, for example, by Ellsberg situations, but I am not convinced that it is true for the Allais type of findings. Assuming that the decision maker did not employ one and the same utility function in the two decision situations in an Allais problem readily explains the observed behaviour. This presumptions, which is both intuitively appealing and highly reasonable, could save the independence axiom and the traditional theory of utility. Thus, had it not been for a number of other experimental results which independently of the Allais experiment also indicate that people do not accept the independence axiom, it may have been premature to say that we have obtained empirical data which call for a search for a new foundation of utility theory.

This list of criteria could be longer and more detailed. However, I think that in the present context these four criteria serve their purpose. Another aspect of empirical robustness which is important is that we tacitly assume that the experiment is carefully designed and carried out by a craftsman (one often forgets that there are few things as difficult as planning and performing a successful psychological experiment). Put somewhat differently, we want our evidence to be obtained by a reliable process, but we also want the object of our investigation to be working properly. Physicists, for example, do not have this problem. They can generally assume that their transmitter (the "world") is working properly; the problem for them is the receptor and the noise that may enter into the measuring procedure. Psychologists, on the other hand, are working under far more complicated experimental conditions. Here both the transmitter, the mediator, and the receptor may be deceptive. The subjects may, consciously or unconsciously deceive us, the results may be caused by a defective experimental design, and due to personal idiosyncrasies we may misinterpret the findings.

Criteria in some respects similar to those above can be put forth for what I would like to call *robustness of the theoretical arguments*, or simply, theoretical robustness. Since this is a paper about the significance of empirical evidence, I will confine myself to briefly outlining three aspect of theoretical robustness.

Firstly, the arguments against the ruling theory should not be of a kind that can, without much difficulty, be handled by this theory, for example, by a reasonable extra assumption. This criterion is exemplified by the above criticism of the Bayesian assumption that the decision maker's beliefs can be represented by a unique probability measure. In order to fully account for the unreliability of a decision situation the Bayesian theory has to give up some of its fundamental presumptions. This criterion corresponds to the above criterion of lack of alternative explanations of the empirical results within the realm of the theory.

Secondly, the theoretical arguments shall prove to us that the axioms or theory under discussion are not as fundamental as we first thought they were. McClennen's (1983) criticism of the sure-thing principle is a good example of a theoretical argument which satisfies this criterion.

Thirdly, we want our arguments against the ruling theory to point us towards a new theory with at least the same analytic power and empirical resilience as the one it replaces. I believe that the arguments against the traditional Bayesian theory or the traditional theory of utility have produced such theories.

This ends my list of criteria.

We now clearly see why it is that only a few empirical results have been (and should have been) of significance for developments in the foundations of decision theory. It is the empirically robust findings that are of importance. But empirical robustness is on its own not enough; the findings have to be backed up by robust theoretical arguments. The Allais and Ellsberg types of results satisfy both these criteria. They have also had a considerable influence on the developments of the theory. The findings on transitivity and reversal of preference are not empirically robust, but even if they had been, they lack support from theoretically robust arguments. These findings have also had considerably less influence on the theoretical developments in decision theory.

Thus, one thing we learn from the history of decision theory is that neither robust empirical arguments, nor robust theoretical arguments, suffice, on their own, to overthrow a consistent and relatively complete prescriptive (normative) theory. However, concurring empirical and theoretical arguments are very forceful, something which is well-known from the theory of evidence. Two witnesses can on their own produce rather weak evidence against the accused, but, if their testimonies are concurring, even the weakest testimony can become forceful and lead to a conviction (see e.g. Sahlin, 1986).

Finally, it should be emphasized that these criteria of empirical and theoretical robustness could, of course, be used to analyze findings also within other research areas where prescriptive models are formulated, i.e. what has been said above is not an isolated decision theoretical phenomenon:

Department of Philosophy
Lund University

NOTE

* The author wishes to thank Peter Gärdenfors, Sören Halldén, Bengt Hansson and William L. Harper for helpful suggestions and discussion.

REFERENCES

Ellsberg, D.: "Risk, Ambiguity, and the Savage Axioms," *Quarterly Journal of Economics*, 71 (1961), 643-669. Reprinted in Gärdenfors & Sahlin (1987).

Gärdenfors, P. and Sahlin, N.-E.: "Unreliable probabilities, risk taking, and decision making," *Synthese*, 53 (1982), 361-386. Reprinted in Gärdenfors & Sahlin (1987).

Gärdenfors, P. and Sahlin, N.-E.: "Decision making with unreliable probabilities," *The British Journal of Mathematical and Statistical Psychology*, 36 (1983), 240-251.

Gärdenfors, P. and Sahlin, N.-E.: *Decision, Probability and Utility: Selected Readings*, Cambridge University Press, forthcoming 1987.

Jeffrey, R. C.: *The Logic of Decision*, McGraw-Hill, New York, 1965; 2nd revised edition, University of Chicago Press, Chicago, 1983.

Kahneman, D. and Tversky, A.: "Prospect theory: An analysis of decision under risk," *Econometrica*, 47 (1979), 263-291. Reprinted in Gärdenfors & Sahlin (1987).
Kahneman, D., Slovic, P. and Tversky, A. (eds.): *Judgment Under Uncertainty: Heuristics and Biases*, Cambridge University Press, New York, 1982.
Kyburg, H. E.: "Bets and beliefs," *American Philosophical Quarterly*, 5 (1961), 54-63. Reprinted in Gärdenfors & Sahlin (1987).
Kyburg, H. E.: "Rational belief," *The Behavioral and Brain Sciences*, 6 (1983), 231-273.
Levi, I.: "On indeterminate probabilities," *Journal of Philosophy* 71 (1974), 391-418. Reprinted in Gärdenfors & Sahlin (1987).
Levi, I.: *The Enterprise of Knowledge*, MIT Press, Cambridge, Mass., 1980.
Lichtenstein, S., and Slovic, P.: "Reversals of preference between bids and choices in gambling decisions," *Journal of Experimental Psychology*, 89 (1971), 46-55.
Machina, M.: "'Expected utility' analysis without the independence axiom," *Econometrica*, 50 (1982), 277-323.
Machina, M.: "Generalized expected utility analysis and the nature of observed violations of the independence axiom," in *Foundations of Utility and Risk Theory with Applications*, B. P. Stigum and F. Wenstøp, Reidel, Dordrecht, 1983, 263-293. Reprinted in Gärdenfors & Sahlin (1987).
McClennen, E.: "Sure-thing doubts," in *Foundations of Utility and Risk Theory with Applications*, ed. by B. P. Stigum and F. Wenstøp, Reidel, Dordrecht, 1983, 117-136. Reprinted in Gärdenfors & Sahlin (1987).
Ramsey, F. P.: "Truth and Probability," in *The Foundations of Mathematics and Other Logical Essays*, ed. by R. B. Braithwaite, Routledge and Kegan Paul, London, 1931, 156-198 and in *Foundations: Essays in Philosophy, Mathematics and Economics*, ed. by D. H. Mellor, Routledge and Kegan Paul, London, 1978, 58-100. Reprinted in Gärdenfors & Sahlin (1987).
Sahlin, N.-E.: "On second order probabilities and the notion of epistemic risk," in *Foundations of Utility and Risk Theory with Applications*, B. P. Stigum and F. Wenstøp, Reidel, Dordrecht, 1983, 95-104.
Sahlin, N.-E.: "Three decision rules for generalized probability representation," *The Behavioral and Brain Sciences*, 8 (1985), 751-753.
Sahlin, N.-E.: "'How to be 100% certain 99.5% of the time'," *Journal of Philosophy*, 83 (1986), 91-111.
Savage, L. J.: *The Foundations of Statistics*, John Wiley, New York, 1954; 2nd revised edition, Dover, New York, 1972.

Slovic, P. and Tversky, A.: "Who accepts Savage's axiom?," *Behavioral Sciences*, 19 (1974), 368–373.

Suppes, P.: "The role of subjective probability and utility in decision-making," in *Proceedings from 3rd Berkeley Symposium on Mathematics, Statistics and Probability*, ed. by I. Neyman, University of California Press, Berkeley, 1956, 61–73.

Tversky, A.: "Intransitive preferences," *Psychological Review*, 76 (1969), 31–48.

EDWARD ERWIN

TESTING FREUDIAN HYPOTHESES

I will discuss three questions: (1) Are *satisfactory* experimental tests of Freudian theory possible? (2) What is the status of the current Freudian experimental evidence? (3) Can Freud's theoretical or therapeutic hypotheses be confirmed without experimental tests?

PART I

Why believe that Freudian theory, or any part of it, is untestable in principle, i.e., that adequate tests are logically impossible? Notturno and McHugh (1986) cite two standard reasons: (1) the theory conflicts with no possible observations; and (2) the theory's proponents refuse to acknowledge falsifications.

The reply to the first argument is well known, and requires little discussion. A particular part of Freud's theory may not *by itself* entail any observation statement, but that is true of most theoretical hypotheses, including many that are clearly testable. To clash with some possible observations, it is sufficient that an hypothesis do so *when combined with auxiliary assumptions*. If, however, auxiliary assumptions may be employed, then it is not obvious that *any* Freudian hypothesis lacks the capacity to clash with some possible observations. One might try to demonstrate for some particular Freudian hypothesis H, that there is *no* consistent set of auxiliary assumptions that can be combined with it to generate an observation statement, but Notturno and McHugh provide no such argument and, as far as I can ascertain, neither has anyone else.

Although Notturno and McHugh mention the first argument for untestability, they may not have wanted to place any weight on it. They rely only on the second argument. However, this too has a standard reply. If an experimental result falsifies a Freudian hypothesis, then it does so whether or not any Freudian concurs. Freudians might reject the falsification for any number of reasons: they might fail to understand the statistics, or the logic of the experimental design, or they may be unfamiliar with the

D. Batens and J. P. van Bendegem (eds.), Theory and Experiment, 123–140.

evidence for one of the auxiliary assumptions employed in the test. Someone who disagrees for any of these reasons may simply be mistaken. In any event, whatever the reason for the disagreement, the mere refusal to accept a disconfirmation is hardly decisive, and in many cases is irrelevant to the issue of whether the apparent counter-evidence is genuine.

Instead of merely balking at the counter-evidence, the Freudians may, as Notturno and McHugh put it (p. 250), use "evasive tactics". For example, they may change their original hypothesis. However, that move does not protect *it* (i.e. the original hypothesis) from the charge of falsification. They may also bring in "ad hoc" assumptions, i.e. assumptions brought in after the experiment has been completed, and use them to explain away the apparent disconfirmation. However, if the invocation of these assumptions is warranted, then the "disconfirmation" was illusory, at least relative to the new evidence offered by the ad hoc assumptions. On the other hand, the ad hoc assumptions may fail to explain away the data in a satisfactory manner; but, then, the disconfirmation stands whether or not the Freudian agrees. On neither alternative, does his or her failure to accept the counter evidence, by itself, cancel the refutation.

At this point in the discussion, a third argument for untestability is likely to emerge. The problem, it might be said, is not that Freudians never give up, but that, given the nature of their theory they always have warrant for resisting putative falsifications (or, more modestly, they have such warrant with respect to certain parts of the theory). In a recent paper defending Popper's views, Cioffi alludes to this third argument when he writes: "Freud often puzzles us, not by putting forward claims without falsifiers but by putting forward other claims as well whose natural force is to cancel the falsifiers of the more straight forward ones" (1985, p. 83). I will refer to this third argument as "the cancellation argument".

In Karl Popper's recent (1986) comments on the testability of Freudian theory, he seems to rely on the first argument for untestability when he says that the theory makes no predictions about overt behavior and, therefore, is untestable. I think, however, that his other remarks suggest that he is using the same argument as Cioffi: the cancellation argument. I do not think he means to deny that one could set up a clash between some description of overt behavior and some particular Freudian theoretical hypothesis by combining the latter with suitable auxiliary assumptions. I think, rather, that his point is this: if the proposition about overt behavior is discovered to be false, one can always appeal to some other part of Freudian theory to explain away the apparent disconfirmation. To use Popper's example, one can employ an hypothesis about a man's unresolved oedipal complex to predict that he will not save a certain child who is drowning, but if he makes an effort to save the child that can also be explained in terms of sublimation.

I find this third untestability argument more interesting than the first two, and will discuss it in the context of a single experiment, one discussed by both critics and friends of Freudian theory.

In 1957, Scodel tested Freud's *orality* hypothesis, which says that male preference for large female breasts results from earlier frustration of oral dependency. He tested the hypothesis by deriving from it and other assumptions the following prediction: Men preferring large breasted women will tend to show more dependency as measured by the Thematic Apperception Test (the TAT). Scodel then performed an experiment and found the opposite result. Men in the small breast preference group gave significantly more TAT dependency themes than both those in the large breast preference group and the no preference group. Scodel concluded that his results were contrary to a widely-held Freudian hypothesis.

In 1972, a Freudian, Paul Kline, considered Scodel's negative results and reacted in exactly the way Popper and Cioffi would predict. He tried to save the orality hypothesis from disconfirmation by postulating reaction-formation in Scodel's small breast preference group. In fact, he suggests (Kline, 1972, p. 91) that Scodel's results can be taken as *support* for Freud's orality hypothesis on the grounds that only Freudian theory could have explained the experimental observations.

In his (1979) discussion, Grünbaum argues that Freudian theory does not necessarily endorse but does *permit* the invocation of reaction-formation to explain Scodel's results. He asks, however, why this permissibility is more damaging to Freudian theory than is a similar permissibility granted by Newtonian theory. The latter theory permits the *ad hoc* postulation of as yet unknown dynamical agencies which could rescue the second law of motion from apparent disconfirmation.

Grünbaum's reply presents a challenge to someone who concludes that Freudian theory is unscientific merely because it permits an *ad hoc* rescue of one of its assumptions by appeal to another part of the theory. His reply, however, leaves open the possibility that Kline was right in invoking reaction-formation to explain why Scodel's results were not really disconfirmatory. If Kline is right, then the Scodel experiment does seem to illustrate nicely the Cioffi-Popper thesis: that for many of Freud's central theoretical assumptions, apparent disconfirmations can always be cancelled by invoking some other part of the theory. I now want to argue, however, that the appearances are misleading.

If we wanted to know if Scodel's result *did* falsify, the orality hypothesis, we would need more empirical information about, for example, the validity of the TAT. At this point in my discussion, however, I am not interested in the truth or falsity of the orality hypothesis, or whether it was actually tested. I am interested in whether it *could* have been tested by Scodel's experiment, or one very much like it. We are free, then, to stipulate that certain conditions *might* have obtained whether or not they were in fact present. For example, assume that prior to Scodel's experiment, there is no firm evidence for Freud's orality hypothesis and that there is solid evidence that reinforcement mechanisms can plausibly explain the type of behavior that Scodel's subjects exhibited. Suppose, further, that we were to find a behavioral correlate of the unconscious anxiety associated with reaction formation. We might find, for example, that among adults who display extreme affection for the parent of the same sex, many report having the same type of recurring, bizarre dream. The most plausible explanation of this finding might be that these people are expressing the "opposite" kind of emotion they consciously feel; unconsciously, they hate their same-sexed parent. This explanation might become more plausible if, after psychoanalytic treatment, such people tended to display hatred toward the parent when and only when the bizarre type of dreaming ceased. Once behavioral indicators of reaction-formation were established in this way, the continued failure to find them in a given case might be evidence of the absence of this particular defense mechanism in a particular individual. If only such individuals were studied in a Scodel-type experiment, then the appeal to reaction formation would not be a defensible explanation of the experimental results.

Given the stipulated background evidence, the reinforcement hypothesis would provide a simpler, more initially plausible explanation of the Scodel results than would be orality hypothesis. The results, then, would clearly favor the reinforcement hypothesis and would also provide *some* (not necessarily overwhelming) evidence against the orality hypothesis.

The Scodel experiment is only one experiment but it illustrates something important, something that runs exactly counter to the Cioffi-Popper thesis. No doubt Freudian defense mechanisms can often be invoked to explain away apparently disconfirming results, but what is doubtful is that such explanations will always be equal to or superior to their rivals no matter what empirical data are obtained. In the Scodel case, if our background evidence was as I stipulated it to be, then Kline's appeal to reaction formation would have been a demonstrable failure; his explanation would have been markedly inferior to Scodel's. In order to support their thesis, Popper and Cioffi have to do far more than merely point out that certain Freudian hypotheses can always be brought in to cancel apparent falsifications. They have to show that the attempted cancellations will always be successful no matter what the experimental evidence may be. They do not argue for this assumption, and neither has anyone else. Consequently, the third argument for untestability fails to show that even one Freudian hypothesis is untestable.

I have now reviewed what I believe to be the three most important reasons for thinking that parts of Freudian theory are untestable. I have not given any positive argument for testability, although in a forthcoming book on psychoanalysis I try to show that *all* of Freud's theoretical hypotheses are testable. Here I will be content with a weaker conclusion: Assuming that there are no better grounds than the three I have discussed, we are not warranted in thinking that any part of Freudian theory is untestable. Why is this important? The reason is this: After arguing on mainly *a priori* grounds that all or part of the theory is untestable in principle, some philosophers and psychologists have concluded that any putative supporting or disconfirming evidence *must* be unsatisfactory. If that were true, then we might have license to speculate about the truth of what Freud said without taking the trouble to examine the experimental or clinical evidence. Such wholesale, *a priori* dismissals of the evidence, however, are unwarranted. If we want to know exactly which of Freud's hypotheses are true, then we cannot reasonably ignore the existing evidence. I turn now to the experimental evidence.

PART II

Forming a reasoned opinion about the status of the Freudian experimental evidence is difficult partly because of the sheer bulk of it. There are now over 1000 experimental studies purporting to test Freudian hypotheses. The reviewer's task has been made much easier, however, because of the work of Paul Kline, and Seymour Fisher and Roger Greenberg. In 1972, Kline reviewed virtually all of the published experimental work on Freudian theory and therapy, and in his 1981 edition, he adds newer studies to his survey. Fisher and Greenberg (1985) review much of the same material and reach conclusions very similar to those of Kline. In his earlier edition, Kline argued that 15 distinctively Freudian hypotheses had been verified (1972, p. 350). In his second edition, he adds an additional hypothesis said to be confirmed by the work of Silverman and his colleagues on subliminal perception. Fisher and Greenberg find confirmation for seven categories of Freudian propositions. Some of these concern the oral and anal character, the role of oedipal and castration factors in male personality development, the etiology of paranoid delusions, the venting function of dreams, and the etiology of homosexuality. The one major disagreement between Kline and Fisher and Greenberg concerns this last hypothesis. Fisher and Greenberg claim confirmation for part of Freud's theory about how homosexuality develops; Kline (1981) criticizes the studies they rely on. Otherwise, Kline and Fisher and Greenberg tend to agree.

My view of the Freudian experimental evidence is more skeptical than that of Kline and Fisher and Greenberg, but there is no space here to comment on each and every study that they find confirmatory. Rather than do that, I will try to categorize the epistemic defects present in a large majority of the studies and will then discuss some important illustrations of these difficulties.

For those experimental studies that are neither confirmatory nor disconfirmatory to any appreciable degree, what are their major problems? The problems fall into several categories. In some cases, one or more of the auxiliary assumptions required for deriving the experimental predictions are either demonstrably false or lack any empirical support. A typical example is the assumption that a certain projective test, such as the Rorschach test, measures accurately the dependent variable. In some cases, the experimenter appeals to Freudian theory to justify reliance on the projective test, but that is not sufficient if the relevant part of Freudian theory also lacks empirical support. Because Kline believes that the validity of the Rorschach test has not been established, he is generally critical of experiments that rely on this test.

A second problem concerns the failure to rule out plausible, alternative explanations of their findings. Some Freudians (e.g., Hall, 1963) take the position that this is unnecessary: if Freudian theory plus warranted auxiliary assumptions imply a prediction, the confirmation of the prediction is sufficient for confirmation of Freudian theory. Kline does not accept this position, and is generally critical of a Freudian interpretation of experimental results where he sees another, plausible interpretation that fits the data just as well. It seems to me that Kline is correct here. There is no need for an experimenter to try to rule out very many of the infinite number of hypotheses that fit his or her results; the background evidence will do that for most such rival hypotheses. However, if a Freudian hypothesis H_1 has an incompatible rival H_2 that explains the data just as well, and is of equal or greater plausibility all things considered, then we are not warranted in believing the Freudian hypothesis on the basis of that experimental evidence. It could be argued that some small degree of confirmation is provided for both H_1 and H_2, but the confirmation, if it exists at all, cannot be of a strong kind. If I know that H_1 and H_2 are incompatible, and that the experimental and background evidence fails to give the edge to either, then I am not warranted in believing one or the other; I may, of course, be warranted in believing the disjunction: either H_1 is true or H_2 is true.

A third problem with some experiments is that the hypothesis they test is not truly Freudian. Sometimes an experimenter will give an operational definition of what is said to be a Freudian concept, but one that drains it of all Freudian or psychoanalytic content. This is one reason why some Freudians are generally skeptical of experimental studies of their theory, and argue for the epistemological priority of the clinical evidence.

One additional problem with the experimental literature should be mentioned. Many of the experiments with reasonably adequate designs need to be replicated. That is especially true where independent investigators have attempted to replicate certain findings, but were unable to do so.

To sum up, the problems with many of the Freudian experiments include the following: the use of unwarranted auxiliary assumptions, the failure to rule out *plausible* rival hypotheses, the testing of hypotheses that are not Freudian, and the failure, where necessary, to replicate the experimental findings.

I will now illustrate the above problems by discussing what are said to be some of the most supportive experimental studies of Freudian theory.

One group of studies, those associated with Lloyd Silverman, illustrates several of the problems I have mentioned. Silverman and his colleagues have conducted a number of interesting experiments using subliminal stimulation to stir up oedipal fantasies. A tachistoscope is used to present negative stimuli to the treatment group, consisting sometimes of pictures, such as one of a snarling man; more often, negative messages are presented to the treatment group. A control group is then shown subliminal neutral pictures of messages, such as the message "People Walking". Silverman claims to have produced some very interesting effects by the use of his technique, including an increase in psychopathology in schizophrenic subjects, an intensification of homosexual orientation in homosexual subjects, and the enhancement of the effectiveness of a behavior therapy technique, systematic desensitization.

One problem with some of Silverman's experiments is his use of questionable tests, such as the Rorschach test, to measure psychopathology or homosexual orientation. To avoid this problem, Silverman, Adler and Lustig (1978) used a simple dependent variable, "competitive performance". They measured this variable in terms of scores in a dart tournament, and claim to have increased the scores of their subjects. An independent investigator, Heilbrun (1980), reports on three experiments in which an attempt was made to replicate the findings. All three attempts failed.

In another study (Silverman et al., 1974), Silverman claims to have enhanced the effectiveness of systematic desensitization by activating unconscious fantasies. However, the experiment failed to rule out a plausible alternative hypothesis: that what Silverman compared was merely a relevant stimulus for the treatment group and an irrelevant one for the control group. To discount this possibility, Emmelkamp and Straatman (1976) used the Silverman experimental design, but used a relevant stimulus for the control group as well. They were unable to replicate the Silverman findings. More recently, Condor and Allen (1980) have also tried to replicate Silverman's study and failed to obtain his results. These failures at replication by independent investigators are especially important given that the positive findings from the Silverman technique have mainly come from experiments done by him, his students or his colleagues. However, these experiments are still being conducted and it is possible that new findings will provide solid evidence for Silverman's hypotheses. There remains at least one problem, however, that is difficult to overcome. Silverman relies on psychoanalytic theory to justify his assumption that his subliminal stimuli stir up unconscious fantasies; no firm independent evidence justifies this assumption. Without such evidence, what is at most demonstrated is that certain kinds of subliminal stimuli produce certain sorts of interesting effects. It would still have to be argued that this has some bearing on psychoanalytic theory.

Another example of a series of studies on subliminal perception is the set of experiments performed by Kragh and his colleagues. In a typical study, Kragh (1960) uses a tachistoscope to present what are called "DMT" and "MCT" pictures to groups of subjects at increasingly greater exposure times. One picture shows a boy with a violin, the head and shoulders of a threatening and ugly male having been inserted at the right of the boy. Another picture shows a young man in the center and an old, ugly man above him. The subjects are instructed to make a drawing of what they have been shown subliminally, without paying any attention to whether their impression is correct. Results are scored using Freudian defence categories. For example, a drawing is classified as "repression" if the main or secondary figure as drawn has the quality of stiffness or rigidity, or is disguised or seen as an animal. What evidence does Kragh offer that repressed wishes are the cause of the subject's drawing the figures in this way? None whatsoever. He simply stipulates that "repression" and other Freudian terms will be applied to certain kinds of drawings, regardless of what caused the subject to make that kind of drawing.

Kline (1981, p. 234) cites Kragh's work as providing strong evidence for the existence of repression, but he concedes that the only evidence for the validity of the DMT and MCT defense measures is what he calls "face validity". However, it is hardly self-evident that these tests really do measure repression; without some supporting evidence for that assumption, the Kragh studies fail to provide any firm evidence for repression, or the operation of any other Freudian defense mechanism.

Examples of studies that confirm propositions that are not distinctively Freudian are those cited by Fisher and Greenberg in support of the anal character. The studies in question do not confirm that people with some so-called "anal" characteristics will display the rest of the syndrome; more importantly, as Fisher and Greenberg concede, they fail to confirm any of Freud's theoretical propositions about the etiology of anal traits or their role in the development of neurosis. What is confirmed is what a non-Freudian could reasonably agree to: that there are at least some people that are parsimonious, compulsive and stubbornly resistant. In commenting on the same evidence, Kline concludes (1972, p. 347) that because there has been a failure to confirm a linkage of so-called "anal" or "oral" traits to infant rearing procedures or to pre-genital eroticism, it would be reasonable to abandon altogether the terms "oral and anal characters".

One final illustration that has received much discussion illustrates the failure to rule out a plausible non-Freudian explanation of the findings. Zamansky (1958) tested five hypotheses that he claimed to be reasonable if one initially assumes the Freudian thesis that paranoids are characterized by strong homosexual desires. The hypotheses were tested by presenting in a tachistoscope-like viewing apparatus pairs of pictures (e.g., of males and females, and of scenes with and without 'homosexually threatening' items). Two of the five hypotheses were not supported, which could be taken as evidence against the initial assumption if Zamansky's other assumptions are all warranted; if the conjunction of his five hypotheses is reasonable given the initial assumption that paranoids are characterized by strong homosexual needs, then the falsification of even one conjunct would falsify the assumption.

However, not all of Zamansky's other assumptions are warranted. One crucial assumption, for which he provides no evidence, is that if the subjects look longer at pictures of males than females when the subject's task is disguised (they were told to determine which picture in each pair was larger), that is because the subjects are repressed homosexuals. Another unwarranted assumption is that when the question of preference of male or female pictures is made explicit, and so more conscious, unconscious defensive forces are set into motion and this causes the preference for male over female pictures is made explicit, and so more conscious, unconscious defensive forces are set into motion and this causes the preferences of the paranoids to approximate those of non-paranoid persons. The failure to provide evidence for either assumption renders Zamansky's results neither confirmatory nor disconfirmatory for the hypothesis that all or most paranoids are repressed homosexuals.

Why did Zamansky's paranoid subjects look longer at the male photos? One cannot be sure. But one plausible explanation, suggested by Eysenck and Wilson (1973), is that being generally suspicious and finding males more of a possible threat, paranoid subjects are likely to pay more attention to pictures of males. Why, however, did the paranoids state a greater preference for pictures of women than men? Eysenck and Wilson suggest that the subjects, after being shown pictures of homosexual encounters, were worried about being labeled 'homosexuals'. However, even this relatively straightforward explanation is not necessary; perhaps the paranoids truly preferred the pictures of females.

In this paper, I have only given illustrations of studies cited by Kline, or Fisher and Greenberg, as being confirmatory, and explained why I disagree. However, in a paper in the Eysenck volume (Modgil and Modgil, 1986), I have examined all of what Kline calls the "exemplary" supporting studies and virtually all of the published studies cited as confirmatory by Fisher and Greenberg. My conclusion there is that the experimental evidence fails to provide strong support for *any* part of Freudian theory. I did not address the issue of whether any part of the theory has been disconfirmed, but even my conclusion about a lack of evidence is controversial and is likely to be challenged. Fisher and Greenberg have not had a chance to comment, but Kline (1986) was able to offer a brief reply in the Eysenck volume. One of his points is that in some cases, I challenge the validity of some projective measure but fail to provide any plausible non-Freudian explanation of the experimental findings. For example, I challenge Friedman's (1952) argument for the existence of a castration complex on the grounds that he assumes, without evidence, that if a subject is told a story about a broken toy elephant and the subject mentions a loss of a tail, then this is good grounds for saying that he has a castration complex. Kline admits that there is no evidence for Friedman's assumption,but challenges me to explain the subjects' responses. I think that can be done without appealing to Freudian theory, but there is a general, philosophical issue that I want to address. I agree with those philosophers of science who argue that an inference to the best explanation is warranted under the following conditions: The explanation is causal; it is superior to those rivals that have been considered; and we have good grounds for thinking that the set of hypotheses that have been considered contains the correct explanation of the data. There may be other conditions as well which warrant an inference to the best explanation, but it is not enough that we have been able to think of only one hypothesis. There are many areas in psychology containing unexplained data; the mere fact that someone provides an explanation, no matter how implausible, is not sufficient grounds for concluding that the explanation is correct. If it were, than anytime someone provided a religious explanation of a currently unexplained so-called "miracle", we would be entitled to infer the existence of a deity.

A second point that Kline makes concerns my objection that some of the hypotheses he cites are not part of Freudian theory. He replies concerning Dixon's (1971) work on subliminal perception, that one must make a value judgment as to whether it confirmed a *Freudian* hypothesis. I do not agree about the particular example, and Dixon himself (1971, p. 244) rejects the idea that his work confirms any Freudian hypothesis. However, I agree with Kline's general point that in some cases one can reasonably disagree as to whether the hypothesis is part of Freudian theory. In such cases, it may be better to ask exactly what hypothesis was confirmed and to leave open the issue of whether that hypothesis is Freudian or not.

Kline makes a similar point about Kragh's work, but again I do not agree about the example. Kline does not want to say that a defense mechanism *caused* Kragh's subjects to draw the way they did, but that, for example, when a subject draws a horrible face as a smiling, pleasant angel, that this is an example of a reaction-formation. In reply, I would agree that Kline is free to stipulate that making such a drawing is to exhibit "reaction-formation", but then "reaction-formation" does not mean what it does in Freudian theory. In postulating reaction-formation, Freud was trying to explain certain behavior patterns; if one refuses to make any causal claim, but simply uses a Freudian expression to *describe* certain observable behavior, then no theoretical claim is being made, Freudian or otherwise.

One final point concerns Eysenck and Wilson's alternative explanation of Zamansky's findings. Kline objects that the explanation is *ad hoc*, embedded in no theory and lacking in intuitive plausibility. Their explanation stands in contrast to the Freudian one which is part of a relatively unified theory.

There are several issues here. First, does Freudian theory have a high degree of unity? Kline's agreement that so-called "Freudian theory" is really a collection of separate mini-theories tacitly concedes that it does not. Second, even if one of these theories, e.g., Freud's account of the etiology of paranoia, is highly unified, this feature is but one explanatory virtue; it might be offset by a lack of initial credibility. We cannot, then, refuse to countenance an alternative, common sense explanation merely because it is not part of a unified theory. The common sense explanation might be more plausible than the Freudian hypothesis despite the latter's being embedded in a unified theory.

Finally, is the Eysenck-Wilson explanation obviously less plausible than its Freudian rival? The former has two parts. First, regardless of which pictures they found more attractive, the paranoids, being more suspicious, tended to fixate more than the non-paranoids on pictures of men because they viewed the men as more threatening than the women. This is somewhat speculative without some evidence that there is a correlation between suspicion and length of looking at threatening figures. However, the Freudian explanation is also speculative, even more so. It requires us to believe that subjects who had no tendency whatsoever to engage in overt homosexual behavior or conscious homosexual fantasizing are nevertheless homosexuals, and that their latent homosexual preferences caused them to stare longer at the pictures of males. Indeed, if the subjects had evidenced homosexual tendencies in any way, that would have refuted Freud's theory that paranoia is caused partly by *unconscious* homosexuality.

The second part of Eysenck and Wilson's explanation is more speculative than the first; it says that relative to non-paranoids, the paranoids expressed a greater verbal preference for the female pictures because they were wary of being labelled a "homosexual". This part of the Eysenck-Wilson explanation, however, is totally unnecessary. What Eysenck and Wilson and Kline all overlook is that there was no statistical difference in verbal preference for the female pictures between the paranoids and non-paranoids. The second alleged positive finding, which Zamansky himself does not claim, was not a finding at all.

So far, I have been discussing Freud's *theory*, but what of the experimental studies of Freudian *therapy*? The answer can be brief: there are no such studies. There have, however, been two reasonably well controlled studies of short-term psychoanalytically oriented therapies. One, conducted by Sloane and his colleagues (1975), compared a therapy of this sort with a behavioral therapy. Clients receiving either therapy improved to approximately the same degree, and greater than those in a wait-list control. However, Sloane emphasizes that the analytic therapy was not psychoanalysis and that no inferences from his results can be reasonably be made about the latter. Furthermore, the Sloane study did not use a placebo therapy in the control group; a minimum contact wait-list control was used instead. Hence, a placebo explanation of the therapeutic findings was not ruled out. Finally, the differences between the two treatment groups and the wait list control were not long lasting; they vanished after thirty days.

A second well known controlled study of psychoanalytically oriented psychotherapy, conducted by Brill and his colleagues (1964), compared this sort of treatment to a sugar pill for psychiatric outpatients. Patients in both treatment groups did better than those in a wait list control, but those receiving the sugar pill did just as well as those receiving the psycho-analytically oriented psychotherapy.

If I am right, then, strong experimental support is lacking for both Freudian theoretical and therapeutic claims. Many analysts would not dispute this assessment. As Paul Meehl has recently noted (1983), even the rare analyst who is familiar with the details of the experimental literature is not likely to have been initially convinced by that literature; for most analysts, including Meehl himself, the clinical case studies are of paramount epistemological significance. I turn, then, to brief comments concerning my third question: Can Freud's theoretical or therapeutic hypotheses be confirmed without experimental tests?

PART III

The argument for the necessity of controlled experiments to establish Freudian therapeutic claims is straight-forward and well known. Freudian therapy generally takes at least two years to complete. One possibility that needs to be discounted is that events occurring outside of the therapist's office during the treatment period caused all or most off the patient's improvement. A second plausible hypothesis is that so-called "placebo" factors, such as the client's having confidence in the therapy, accounted for the improvement. There is ample evidence that these two hypotheses, the spontaneous remission and placebo hypotheses, can plausibly account for documented findings of clinical case studies for most psychological therapies. Controlled experiments using a credible placebo, then, are needed to rule out plausible rivals to Freudian therapeutic hypotheses. The epistemic situation does not change appreciably if the analyst claims not to eliminate symptoms but rather to produce insights or some other benefit. If the benefit is non-trivial, we will still generally need controlled studies to establish that the alleged effect did occur and was the result of the therapy.

One recent challenge to the need for controlled study can be dealt with briefly. Some writers (e.g. Cordray and Bootzin, 1983) are suggesting that placebo factors need not be ruled out. They argue that if a patient's faith in the therapy caused all improvement, then this does mean that the therapy was ineffective; rather, it means that a certain psychological explanation of the therapy's effectiveness is correct. The reply is this: if psychoanalysis is not more effective than a minimal placebo, such as the sugar pill used in the Brill study, then serious problems arise for the psychoanalytic enterprise. First, it will be difficult to retain the Freudian view that neurotic symptoms are maintained by deeply rooted repressed wishes. Second, it becomes difficult to defend the training of analysts, or even the use of analysis in cases where an inexpensive placebo will do as well.

A second argument against the need for controlled study has been made recently by Jerome Frank. He argues that the results of meta-analysis show that all psychological therapies are effective, and are effective to approximately the same degree; so, he infers that psychoanalysis is also effective.

What is meta-analysis? It is a relatively new statistical technique introduced to solve an epistemological problem that ought to be of interest to philosophers: the integration problem. The problem is to develop a rule for integration of experimental findings that are disparate or inconsistent. For example, suppose that controlled studies of uneven quality yield conflicting evidence about the effectiveness of a certain kind of therapy. How do we weight the evidence from the various studies? One answer is to use the box-score method. Assign a numerical weight to each study depending on how may defects exist in the experimental design. The box score method, however, encounters several problems; one is that any précise weighting scheme appears arbitrary. Smith, Glass and Miller (1980) offer the following solution. They calculate what they call the "effect size" for a study by subtracting the average score for the control group from the average score for the treatment group and dividing the result by the standard deviation of the control group. If a study has more than one measure of treatment outcome, than a separate effect size is calculated for each measure.

In their book on the benefits of psychotherapy, Smith, Glass and Miller (1980) review virtually all of the controlled studies of psychotherapy, 473 studies in all. For these studies, they found 1760 effect sizes and then calculated the average effect size for psychotherapy. Among other things, they concluded that all forms of psychotherapy, including behavior therapy, produce substantial therapeutic benefits and are equally effective.

The topic of meta-analysis has been of intense interest to psychologists in recent years; at least, four journals have devoted entire issues to the subject. I do not want to argue that this much discussed technique is of no value, but I have recently tried to demonstrate (1984) that Smith and Glass's particular use of it has serious defects which completely undermine their argument for the effectiveness of psychotherapy. The defects concern mainly their giving equal weight to poor and good experimental studies. Apart from these defects, there is another serious problem. As Eysenck (1983) has pointed out, even if their calculations are accepted without criticism, the effect size they list for the psychodynamic psychotherapies are only marginally greater than the effect size for what they call a "placebo" therapy. In other words, even taken at face value, their meta-analyses failed to show that the dynamic psychotherapies are appreciably better than a placebo.

I close with a few comments about the need for controlled study of Freudian theory. Freud himself denied such a need, but also pointed to a grave threat to his interpretation of his clinical findings: that the findings were due to unintended suggestions by the analyst. As Grünbaum notes in his recent book (1984), Freud had a strategy for solving this problem. He relied on what Grünbaum calls the "Tally Argument". Roughly: (1) A necessary condition of therapeutic improvement is that the analyst's interpretations be correct; (2) patients sometimes do show therapeutic improvement; therefore, (3) some of the interpretations of analysts are correct. As Grünbaum points out, however, the first premise of this argument has been disconfirmed by spontaneous remissions and by the demonstrated effectiveness of certain behavior therapies. Grünbaum also goes on to argue that without some alternative decontaminating strategy, the clinical evidence has too many epistemic defects to provide any firm support for Freud's causal hypotheses.

I will not repeat here the details of Grünbaum's arguments, but they are quite powerful. Anyone who believes that clinical evidence can serve as an adequate substitute for experimental evidence needs to answer his arguments.

A number of philosophers and psychoanalysts *have* replied to Grünbaum in a recent issue of the *Behavioral and Brain Sciences;* Grünbaum's counter-replies are also included (Grünbaum, 1986).

One typical reply is to point out that modern psychoanalysts do not rely on Freud's Tally Argument. That reply, however, is of no help at all to the psychoanalyst unless some substitute for the Tally Arguments is also provided. A second standard reply is that modern psychoanalysts no longer rely on Freudian theory. Grünbaum, however, has already anticipated that reply; Non-Freudian versions of psychoanalytic theory must confront the same sorts of epistemological difficulties if they are to be confirmed by clinical, non-experimental evidence. There are, it might be added, virtually no experimental studies of the newer versions of psychoanalytic theory. Finally, an interesting reply to Grünbaum has been developed by the psychoanalyst Marshall Edelson. Edelson (1984) argues that single subject and quasi-experimental designs, which have been used with success in studies of behavior therapy, can be adapted to the Freudian clinical setting. I have grave doubts about this proposal, as does Grünbaum, partly because of the vast differences between the relatively simple behavior therapy hypotheses and the complex Freudian hypotheses that need to be tested. However, whether or not Edelson's proposal can be carried out, that has not, with a few possible exceptions, been accomplished to date. So, even if Edelson's proposal has merit, this does nothing to show that the current clinical evidence supports Freudian theory. Conclusion: I began by arguing that the standard reasons for thinking Freudian theory untestable in principle are inadequate. If I am right, one cannot dismiss Freudian claims without assessing the experimental and clinical evidence. My own assessment of the experimental evidence, argued for only briefly in this paper, is that Freud's theoretical and therapeutic claims all lack strong experimental support. I did not argue, however, that any of Freud's claims are false. Finally, I briefly argued that the current clinical evidence is not likely to confirm Freud's therapeutic claims, nor, if Grünbaum's arguments are sound, does such evidence provide any strong support for Freudian theory.

University of Miami
Coral Gables, Florida

REFERENCES

Brill, N., Koegler, R. Epstein, L. Fogey, E. (1964), Controlled study of psychiatric outpatient treatment. *Archives of General Psychiatry*, 10, 581-595.

Cioffi, F. (1985), Psychoanalysis Pseudo-science and testability. In G. Currie and A. Musgrave (eds.), Boston; Martin Nijhoff.

Condor, T. & Allen, G. (1980), Role of psychoanalytic merging fantasies in systematic desensitization: A rigorous methodological examination. *Journal of Abnormal Psychology*, 89, 437-443.

Cordray, D. & Bootzin, R. (1983), Placebo control conditions: Tests of theory or of effectiveness? *Behavioral and Brain Sciences,* 6, 286-287.

Dixon, N. (1971), *Subliminal perception: The nature of a controversy.* New York: McGraw Hill.

Edelson, M. (1984), *Hypothesis and evidence in psychoanalysis,* Chicago: University of Chicago Press.

Emmelkamp, P. & Straatman, H. (1976), A psychoanalystic reinterpretation of the effectiveness of systematic desensitization: Fact or fiction? *Behav. Res. & TherapyJ, 14-245-249.*

Erwin, E. (1984) Establishing casual connections: Meta-analysis and psychotherapy. *Midwest Studies in Philosophy.*

Erwin, E. (1986), Psychotherapy and Freudian psychology. In S. Modgil & C. Modgil (Eds.) *Hans Eysenck: A Psychologist Searching for a Scientific Basis for Human Behavior.* Falmer Press.

Eysenck, H.J. (1983), The effectiveness of psychotherapy: The specter at the feast. *Behavioral & Brain Sciences*, 6, 290.

Eysenck, H.J. & Wilson, G. (1973), *The experimental study of Freudian theories.* London: Methuen & Company Ltd.

Fisher, S. & Greenberg, R. (1985), *The scientific credibility of Freud's theories and therapy.* New York: Columbia University Press.

Friedman, S. (1952), An empirical study of the castration and Oedipus complexes. *Genet. Psychol. Monogr.*, 46, 61-130.

Grünbaum, A. (1979), Is Freudian psychoanalytic theory pseudoscientific by Karl Popper's demarcation criterion? *American Philosophical Quarterly,* 16, 131-141.

Grünbaum, A. (1984), *The Foundations of Psychoanalysis: a Philosophical Critique*. Los Angeles: University of California Press.

Grünbaum, A. (1986), Precis of *The Foundations of Psychoanalysis: A Philosophical Critique. Behavioral and Brain Sciences*, 9, 217-228.

Hall, C. (1963), Strangers in dreams: An experimental confirmation of the Oedipus complex. *J. of Pers.*, 31, 336-345.

Kline, P. (1972, 1981), *Fact and fantasy in Freudian theory.* London: Methuen

Kline, P. (1986), Kline replies to Erwin. In S. Modgil & C. Modgil (Eds.) *Hans Eysenck: A Psychologist Searching for a Scientific Basis for Human Behavior.* Falmer Press.

Meehl, P. (1983), Subjectivity in psychoanalytic inference: the nagging persistence of Wilhelm Fliess's Achensee question. In J. Earman (ed.) *Testing Scientific Theories, Minnesota Studies in the Philosophy of Science, vol. X.,* Minneapolis, University of Minnesota Press.

Notturno M. and McHugh P. (1986). Is Freudian psychoanalytic theory really falsifiable? *Behavioral and Brain Sciences*, 9, 250–252.

Popper, K. (1986), Predicting overt behavior versus predicting hidden states. *Behavioral and Brain Sciences,* 9, 254–255.

Scodel, A. (1957), Heterosexual somatic preference and fantasy dependency, *Journal of Consulting Psychology,* 21, 371-374.

Silverman, L., Frank, S. & Dachinger, P. (1974), Psychoanalytic reinterpretation of the effectiveness of systematic desensitization: Experimental data bearing on the role of merging fantasies. *J. of Abnormal Psychol.*, 83, 313-318.

Silverman, L., Ross, D., Adler, J. & Lustig, D. (1978), Simple research paradigm for demonstrating psychodynamic activation: Effects of Oedipal stimuli on dart-throwing accuracy in college males. *J. of Abnormal Psychol.,* 87, 341–357.

Sloane, R., Staples, F., Cristol, A., Yorkston, N. and Whipple, K. (1975) *Psychotherapy versus Behavior Therapy.* Cambridge, Mass.: Harvard University Press.

Smith, M., Glass, G. & Miller, T. (1980), *The benefits of psychotherapy.* Baltimore, Md. : Johns Hopkins University Press.

Zamansky, H. (1958), An investigation of the psychoanalytic theory of paranoid delusions. *J. of Pers.,* 26, 410–425.

ALLAN FRANKLIN

EXPERIMENT, THEORY CHOICE, AND THE DUHEM-QUINE PROBLEM

The fallibility and corrigibility of experimental results is well established. One need only look at a survey of the measurements of fundamental physical constants[1] or at the "Review of Particle Properties,"[2] a standard reference for high-energy physicists, to see that the measured values of not only particular quantities, but also of the world average of such quantities, change by far more than their cited experimental uncertainties would suggest is plausible or probable.[3] A case in point is the history of measurements of η_{+-}, the CP violating parameter in K^0 decay (see Figure 1). Prior to 1973, η_{+-} had been measured six times. The results were in good statistical agreement and had a mean of $(1.95 \pm 0.03)\times 10^{-3}$. The four succeeding measurements agreed with each other, and had a mean of $(2.27 \pm 0.022)\times10^{-3}$. These means differ by eight standard deviations, an unlikely change if these are two sets of correct measurements of the same quantity. (The probability is 1.24×10^{-15}). Unless one is willing to consider the possibility that η_{+-} changed in 1973, we must conclude that at least one of these sets of measurements is incorrect.

This fallibility and corrigibility of experimental results necessarily extends to the confirmation or refutation of theories or hypotheses based on those results. This is nicely illustrated by the history of nuclear β decay during the 1930s. In 1934, Fermi proposed a theory of such decays.[4] His theory agreed with the shape of the observed energy spectra. It also predicted that the quantity $F\tau_0$ would be approximately constant for different types of decay,[5] where F is the integral of the energy distribution, and τ_0 is the lifetime of the transition. This was confirmed by Sargent's previously reported work.[6]

It was quickly pointed out by Konopinski and Uhlenbeck[7] that more detailed examination of the spectra showed discrepancies. Fermi's theory predicted too few low energy electrons and an average decay energy that was too high. They proposed a modification of the theory that remedied these deficiencies. They cited as support for their model the spectra of P^{30}, obtained by Ellis and Henderson,[8] and of RaE, found by Sargent.[9] The experimental spectra, along with the predictions of the two competing theories, are

D. Batens and J. P. van Bendegem (eds.), Theory and Experiment, 141–155.

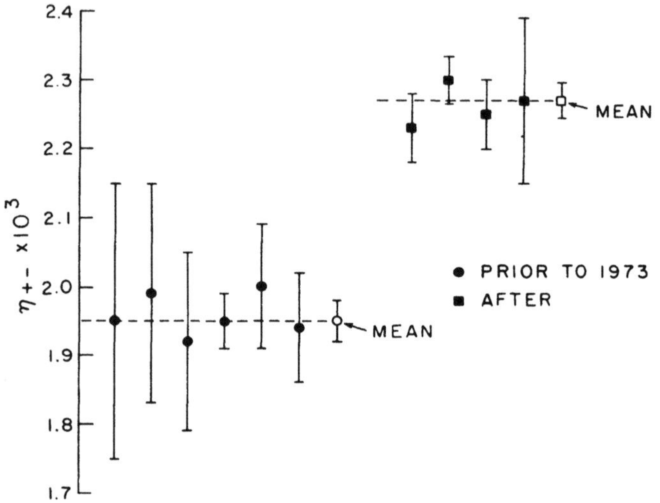

Figure 1. Measurements of η_{+-} in order of their publication.

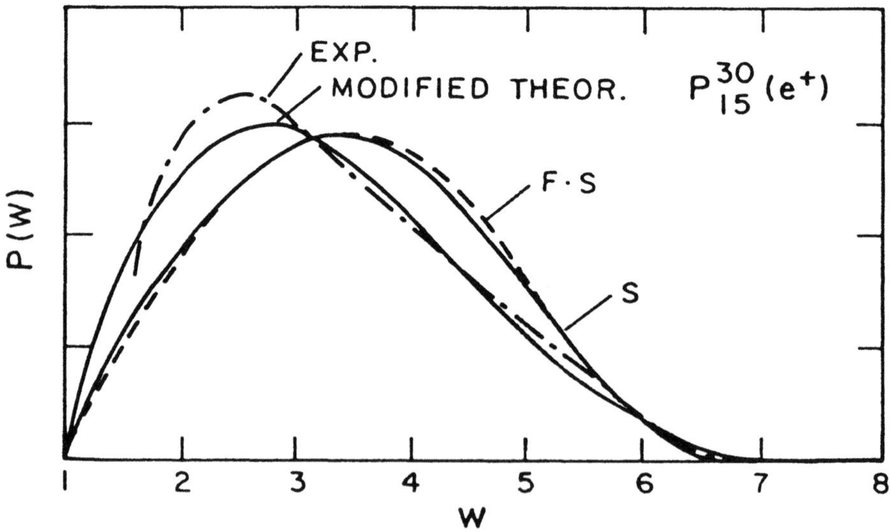

Figure 2. Spectrum of P^{30}. From Konopinski and Uhlenbeck (note 7).

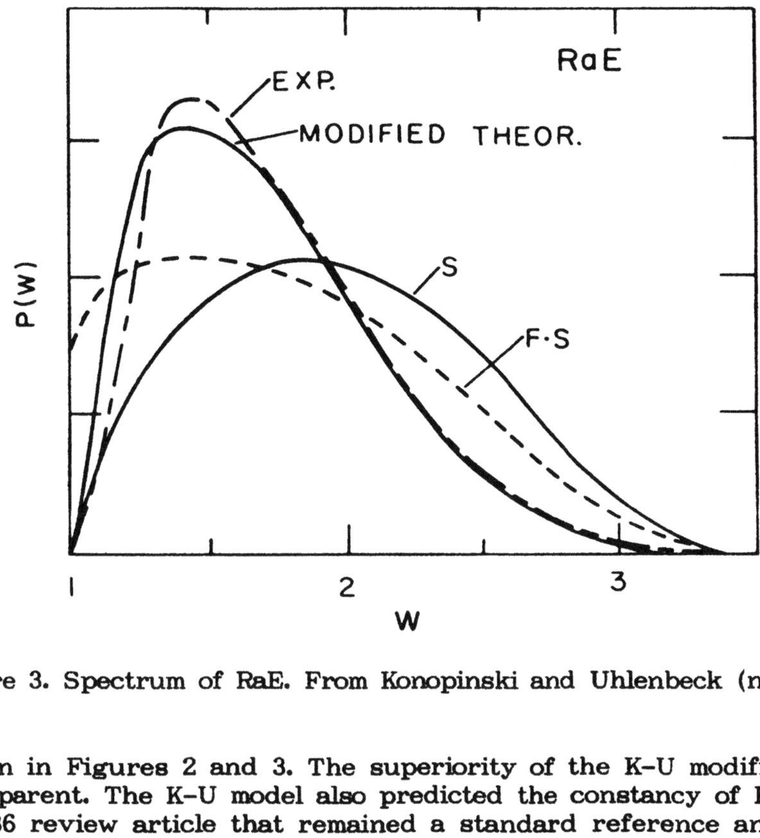

Figure 3. Spectrum of RaE. From Konopinski and Uhlenbeck (note 7).

shown in Figures 2 and 3. The superiority of the K–U modification is apparent. The K–U model also predicted the constancy of $F\tau_0$. In a 1936 review article that remained a standard reference and was used as a student text into the 1950s, Bethe and Bacher remarked, "We shall therefore accept the Konopinski-Uhlenbeck theory as the basis of future discussions."[10]

Further experimental work during the late 1930s and early 1940s showed that the early energy spectra, as well as others that had subsequently provided support for the K–U theory, had systematic experimental problems. Scattering and energy losses in thick sources had resulted in observed spectra with too many low energy electrons and too low an average energy. As sources were made thinner the results approached the predictions of Fermi's original theory (see Figure 4). The correct theory should give a good fit to a straight line.

It was also pointed out that the spectrum calculated by Fermi was for "allowed" transitions, whereas RaE and several of the other elements whose beta decay spectra had provided support for

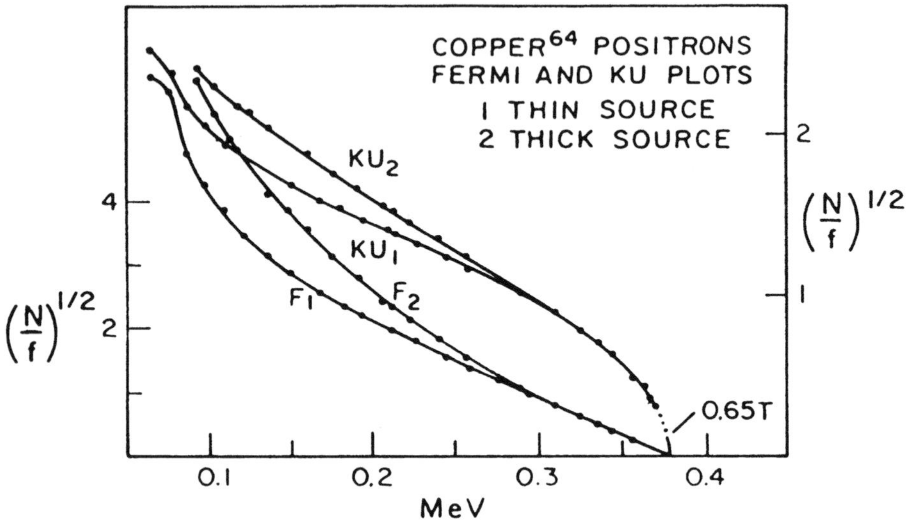

Figure 4. Spectrum of Cu^{64} for both thick and thin sources. The predictions of Fermi and of Konopinski and Uhlenbeck are shown. The correct theory should be a straight line. From A.W. Tyler, "The Beta- and Gamma-Radiations from $Copper^{64}$ and $Europium^{152}$," *Physical Review, 56* (1939), 125–30.

the K–U theory were "forbidden" transitions. This had been mentioned earlier, but not much attention had been paid to it. The predicted spectra for the two types of transition were not expected to be the same. When the Fermi theory spectrum for forbidden transitions was calculated, ironically, by Konopinski and Uhlenbeck,[11] the discrepancy disappeared. The spectrum of phosphorus, P^{32}, is shown in Figure 5.[12] The superior fit of the Fermi theory is clear.

The early spectra confirmed the K–U theory and refuted Fermi's theory. After these experimental results were found to be in error, the new results reversed that decision. In 1943, Konopinski published a comprehensive review of β decay and concluded, "Thus, the evidence of the spectra, which had previously comprised the sole support for the K–U theory now definitely fails to support it."[13]

Thus, we have seen the fallibility and corrigibility of experimental results, of theoretical comparison with data, and of the confirmation and refutation of theories.

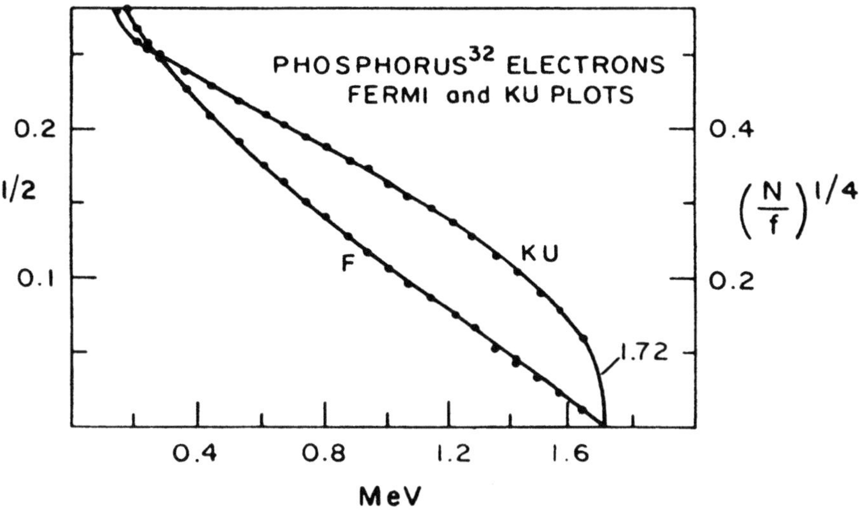

Figure 5. The spectrum of P^{32} showing the predictions of Fermi and of Konopinski and Uhlenbeck. From Lawson (note 12).

Does this episode, along with the general fallibility of experimental results discussed earlier, argue against a legitimate role for experiment in theory choice? I think not. Not all incorrect experimental outcomes result in incorrect theory choices. Thus, both sets of measurements of η_{+-} confirm the hypothesis of CP violation, which requires only that $\eta_{+-} \neq 0$. The fact that Millikan's value of e, the charge on the electron, $(4.774 \pm 0.009)\times 10^{-10}$esu, disagrees with the currently accepted value of $(4.803242 \pm 0.000014)\times 10^{-10}$esu has not changed the support for charge quantization. The early history of β decay shows only that we can be mistaken in our theory choice, and not there were not good reasons for that choice. I will argue later that there are good reasons for rational belief in experimental results and that these results then provide the basis for theory choice. The fact that rational belief does not guarantee truth should not preclude justified and reasonable theory choice.

I have argued elsewhere,[14] in some detail, that there is an epistemology of experiment, a set of strategies that provide rational belief in an experimental result, and which are also used by practicing scientists. These strategies distinguish between a valid observation or measurement and an artifact created by the experimental apparatus.

One such strategy, suggested by Hacking,[15] is intervention. Hacking's example involves the use of the microscope, with the experimenter "intervening" (i.e., staining, injecting fluid), and in other ways manipulating the object under observation. This is a special case of a more general strategy in which we predict what will be observed after the intervention if the apparatus is working properly or as we expect it to. When the observation is made, we strengthen our belief in the proper operation of the apparatus and in its results. Thus, if we inject stain into a cell we expect that its appearance under the microscope will change. When it does, we strengthen our belief in the validity of microscope images.

Hacking also suggests that we believe in an observation if it can be made using "different" microscopes (i.e., optical, polarizing, phase-contrast, etc.) or if it is independently confirmed. This is a special case of the greater confirmation of a hypothesis by "different" experiments than by repetitions of the "same" experiment.[16] At least one reason for calling experiments different is if the theories of the apparatus are different. It is our theory of microscopes and of light that tells us that the various microscopes are different. Theoretical context may also help to determine what is or is not a different experiment. It seems clear, then, that here theory-ladenness is a virtue rather than a defect.

If the apparatus is based on a well-corroborated theory, we also have good reasons to trust its results. The support for the theory feeds through to the observations. Even without such a theory we can validate observations. Sometimes the phenomena themselves provide such evidence. The observations of the moons of Jupiter by Galileo were extremely unlikely to have been artifacts of the telescope, as suggested by Cremonini, because they appeared to represent a consistent planetary system, and even obeyed Kepler's Third Law (R^3/T^2 = constant). The consistency of Millikan's data, in which all observed charges were integral multiples of a minimum unit of charge, helped to validate his results. No remotely plausible malfunction of the apparatus could produce such a consistent result.

If all plausible sources of error and all alternative explanations can be eliminated, then an observation is valid.[17] Thus, when electronic signals were observed by Voyagers 1 and 2, the following possible causes were eliminated: defects in the telemetry, interaction of the spacecraft with the environment of Saturn, lightning, dust, etc. We were left with electric discharges in the rings of Saturn as the cause of the signals.

Perhaps the most widely used strategy for validation of results is that of calibration and experimental checks. The ability of an apparatus to reproduce already known phenomena argues both for its proper operation and in favor of the results obtained. If my spectrometer reproduces the known Balmer series of hydrogen, then I have reason to believe its measurement of the spectrum of an unknown substance. The fact that an observation can be explained using currently accepted theory can also help to provide validation. At least part of the reason why physicists accepted the discovery of the W boson was that its mode of production, decay mode, angular distribution, and mass were in agreement with the independently well-corroborated Weinberg-Salam theory.

I further argued that these strategies have plausible justification and provide us with rational belief in experimental results. These results can then serve as a basis for theory choice.

Some sociologists of science, as well as some philosophers and historians of science, have argued against such a role for experiment. A typical view is given by MacKenzie. "Recent sociology of science, in alliance with sympathetic tendencies in the history and philosophy of science, has shown the inadequacy of any experiment, or set of experiments however large, on its own to compel resolution of a point of controversy, or, more generally, acceptance of a particular fact. A sufficiently determined critic can always find reason to dispute any alleged 'result'. If the point at issue is, say, the validity of a particular theoretical claim, those who wish to contest an experimental proof or disproof of the claim can always point to the multitude of auxiliary hypotheses (for example about the operation of instruments) involved in drawing deductions from a given theoretical statement to a particular experimental situation or situations. One of these auxiliary hypotheses may be at fault, critics can argue, rather than the theoretical claim apparently being tested. Further, the validity of the experimental procedure can also be attacked in infinitely many ways."[18]

There are two points at issue here. The first involves the meaning one assigns to "compel". If one reads it, as MacKenzie seems to, as "entail" then I agree that no finite set of confirming instances can entail a universal statement. No matter how many white swans one sees it does not entail that "all swans are white". A single instance can, however, refute a universal statement. Observation of a single black swan refutes "all swans are white."[19] Less artificial examples will be discussed later. It seems to me, however, that a more plausible meaning for "compel" is having good and rational reasons for belief. As the episodes discussed later demonstrate, this is the meaning used in scientific discussions.

The second point is a logical one, known to philosophers of science as the Duhem–Quine problem,[20] and discussed by MacKenzie. In the usual *modus tollens* if a hypothesis h entails an experimental result e then $\neg e$ (not e) entails $\neg h$. As Duhem and Quine pointed out it is not just h that entails e but rather h and b, where b includes background knowledge and auxiliary hypotheses. Thus, $\neg e$ entails $\neg h$ or $\neg b$ and we don't know where to place the blame. One can, of course, also question the experimental result $\neg e$.

As discussed later, there is at least one example taken from the practice of science, where if one accepts the experimental evidence, one can solve the problem of where to place the blame. More important, however, is the fact that scientists never confront all the logically possible explanations of a given result. There are usually only a reasonable number of plausible or physically interesting alternatives on offer. In this case scientists evaluate the cost of accepting one of these alternatives in the light of all existing evidence. The alternatives themselves may be tested, subject to the usual difficulties, and we may be left with only one explanation. An example of this will also be presented later.

Let us first consider a weak form of the Duhem–Quine problem, in which one accepts the experimental results as correct. There is at least one philosophically legitimate solution to this problem. Consider the discovery of parity nonconservation, or the violation of mirror reflection symmetry in nature.[21] In this case, there are in fact only two classes of theories, those that conserve parity and those that do not. These classes are mutually exclusive and exhaustive. The experimental test was as follows. Suppose we orient the spin of a radioactive nucleus so that it points upward. Also suppose that the electron from the β decay of this nucleus is emitted in the direction opposite to the spin, or downward. As shown in Figure 6 the mirror image of this result differs from the original. In the mirror the electron is emitted along the spin direction. This was the experiment performed by Wu and associates.[22] The results are shown in Figure 7 and clearly show an asymmetry. Once one accepts the evidence in favor of the observed asymmetry then it seems that one must either attribute it directly to nonconservation of parity in the weak interactions or else say that parity is conserved in the weak interactions and modify the background beliefs so that the universe itself is asymmetric. There does not seem to be any alternative that preserves both

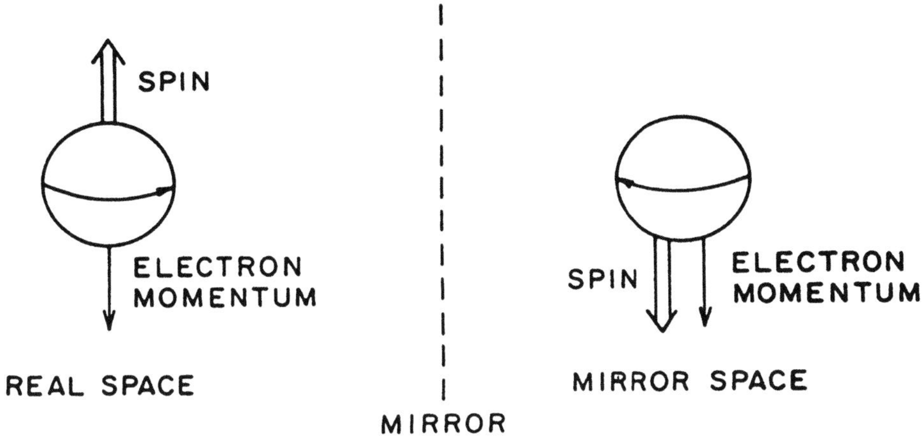

Figure 6. The β decay of oriented nuclei in real and mirror space.

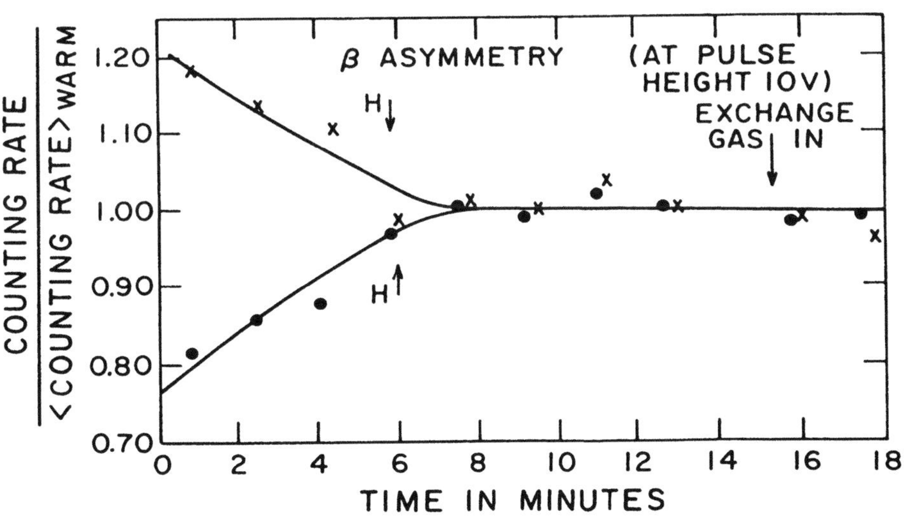

Figure 7. The decay electron counting rate as a function of time. for the decay of polarized Co^{60} nuclei. The asymmetry between the two polarizations is shown. From Wu et al. (note 22).

parity conservation and the symmetry of the universe, given the experimental results. It is the very general nature of symmetry principles like parity conservation that allows them to be refuted. In this case we are compelled to accept the class of theories that do not conserve parity.

A more complex case is the demonstration that the combined symmetry CP (charge-conjugation, or particle-antiparticle symmetry, and parity) was violated.[23] CP conservation allowed the K^0_1 meson, but not the K^0_2 meson, to decay into two pions. The experiment of Christenson and associates[24] found $K^0_2 \rightarrow \pi^+\pi^-$ and thus, in the most obvious interpretation, demonstrated CP violation. This was the view taken in 3 out of 4 theoretical papers written during the period immediately following the report of those results. The rest offered alternative explanations. These relied on one or more of three arguments: (1) the Princeton results are caused by a CP asymmetry (the local preponderance of matter over anti-matter) in the environment of the experiment, (2) $K^0_2 \rightarrow \pi^+\pi^-$ does not necessarily imply CP violation, and (3) the Princeton observations did not arise from $K^0_2 \rightarrow \pi^+\pi^-$ decay. This last argument can be divided into the assertions (3a) the decaying particle was not a K^0_2, (3b) the decay products were not π mesons, and (3c) another unobserved particle came off in the decay. Included in these alternatives were three suggestions that cast doubt on well-corroborated fundamental assumptions of modern physics. These were: (1) pions are not bosons, (2) violation of the principle of superposition in quantum mechanics, and (3) failure of the exponential decay law. Although by the end of 1967 all of these alternatives had been experimentally tested and found wanting, the vast majority of the physics community had accepted CP violation by the end of 1965, even though all the tests had not yet been completed. As Prentki remarked, this was because in some cases "the price one has to pay in order to save CP becomes extremely high," and because other alternatives were "even more unpleasant."[25] This is an example of what one might call a pragmatic solution to the Duhem-Quine problem. Here the alternative explanations, which modified the background knowledge, and the auxiliary hypotheses were refuted, leaving CP unprotected. There were certainly other possible alternative explanations, but none were offered, no doubt because they were not regarded as plausible or physically interesting.[26]

A worrisome question still remains, however. The argument in the case of parity violation, in which the two classes of theories are both exclusive and exhaustive, would seem to apply just as well to the episode of CP violation. After all, there are only two classes of theories: those that conserve CP and those that do not.

What is the difference between the two episodes? I believe the difference lies in the length and complexity of the derivation linking the hypothesis to the experimental result, or to the number of auxiliary hypotheses required. In the case of parity, the experiment could be seen by inspection to violate mirror symmetry (see Figure 6). In the CP episode, what was observed was $K^0_2 \rightarrow \pi^+\pi^-$. In order to relate this observation to the question of CP violation, one had to assume (1) the principle of superposition, (2) that the exponential decay law held to 300 lifetimes, (3) that the decay particles were both "real" pions and that pions were bosons, (4) that no other similar particle was produced, and (5) that there were no external conditions present that might regenerate K^0_1 mesons. This is a far cry from the direct derivation in the parity experiment, and it was these auxiliary assumptions that were tested and eliminated by subsequent experiments.

Let us consider another episode from the history of physics. During the 1930s, Mott, on the basis of Dirac's theory of the electron, calculated that in double scattering of electrons from heavy nuclei (high *Z*, the charge on the nucleus) there would be a forward-backward asymmetry.[27] The experiment was performed many times and the conclusion was that the expected asymmetry had not been observed. Despite this, belief in Dirac's theory remained quite strong. Mott's derivation required several assumptions including single scattering, at high velocity, off high-*Z* nuclei, at large angles and the assumption that other effects such as screening of the nucleus by atomic electrons, multiple scattering, and modifications of Coulomb's Law were negligible. It was the question of whether or not the experimental apparatuses satisfied the conditions for Mott scattering and whether or not these other effects might be present that were raised. Dirac's theory went virtually unquestioned. Why was this? I suggest that it was because of the success of the theory in predicting the existence of the positron, a result totally unexpected on the basis of background knowledge or any other existing theory, and its success in atomic spectroscopy. In comparison with these successes the discrepancy in electron scattering did not carry sufficient weight.[28]

We may clarify this idea of an "important" confirming or refuting experiment, and the dependence of this importance on the theoretical context using Bayesian confirmation theory. Let D be the Dirac theory and e_1 be the existence of the positron. Applying Bayes's Theorem, we get

$$P(D|e_1) = P(e_1|D)P(D)/P(e_1) \text{ where } P(e_1|D) = 1 \text{ because } D \vdash e_1.$$

We can also write

$$P(e_1) = P(e_1|D)P(D) + P(e_1|\neg D)P(\neg D)$$

where $\neg D$ is the negation of D. Because no other theory, at the time, predicted the existence of the positron, we can set $P(e_1|\neg D)$ $P(\neg D) = \beta$, where β is very small. Thus, $P(D|e_1) = P(D)/(P(D)+\beta) \approx 1 - \varepsilon \approx 1$, where $\varepsilon = \beta/P(D)$ is also quite small. This confirms our intuition that observation of a result predicted by only one theory gives us very good reason to believe that theory. This is not, of course, the only type of important experiment. The examples discussed earlier of the experiments on parity nonconservation and CP violation provide an illustration of another type. These are experiments that eliminate competing hypotheses or theories while confirming one of the alternatives. These also provide substantial confirmation.[29]

We now set $P(D|e_1) = P_1(D)$, the new prior probability of D. Let us now consider what happens to our belief in D when e_2, the symmetric double scattering of electrons, is observed.

$$P(D|e_2) = P(e_2|D)P_1(D)/P(e_2) \quad (1)$$

$$P(e_2) = P(e_2|D)P_1(D) + P(e_2|\neg D)P_1(\neg D) \quad (2)$$

Let $K = P(e_2|\neg D)/P(e_2|D)$; then, from equation (2)

$$P(e_2) = P(e_2|D)P_1(D) + K\,P(e_2|D)[1 - P_1(D)]$$

but $P_1(D) = 1 - \varepsilon$

$$P(e_2) = P(e_2|D)[1 - \varepsilon] + K\,P(e_2|D)[1 - 1 + \varepsilon]$$

$$P(e_2) = P(e_2|D)[1 - \varepsilon + K\varepsilon]$$

We can neglect ε, which is very small[30]: so, from equation (1),

$$P(D|e_2) = P(e_2|D)P_1(D)/P(e_2|D)[1 + K\varepsilon] = P_1(D)/[1 + K\varepsilon]$$

We expect $K \leq 1$, because $\neg D$ is a weak statement compared with D and in this case we see $P(D|e_2) \approx P_1(D)$ or our confidence in D remains virtually unchanged. The result is quite robust for differing values of K, depending on ε. Thus if $\varepsilon = 0.01$, K can be as large as ten without changing our belief in D by more than 10 percent.

Let us now see what we can say about $P(H|e_2)$ where H represents the auxiliary hypotheses used by Mott to derive $\neg e_2$, the asymmetric scattering: single, large angle scattering of high velocity electrons from a heavy nucleus. By a theorem of the probability calculus,

$$P(H|e_2) + P(D|e_2) - P(H \wedge D|e_2) = P(H \vee D|e_2)$$

but $P(H \wedge D|e_2) = 0$, because $H \wedge D \vdash \neg e_2$; so

$$P(H|e_2) = P(H \vee D|e_2) - P(D|e_2) \quad (3)$$

But $P(H \vee D|e_2) > P(D|e_2)$, so we can write $P(H \vee D|e_2)$ as $1 - \delta$, where $\delta \leq K\varepsilon$

$$P(H|e_2) = 1 - \delta - (1 - K\varepsilon) = K\varepsilon - \delta,$$ which is very small.

Thus, as both our simple calculation and the history show, our belief in H becomes very low, whereas our belief in D remains virtually the same. Scientists did in fact offer alternatives to H.

This is a solution to a very particular case of the Duhem-Quine problem in which P(D) is originally very large. In the general case, in which $P(D|e_2)$ is not very close to 1, we cannot say anything as equation (3) shows. It does indicate, however, that when $P(H|e_2)$ is close to 1, D will receive the blame, as we expect.

This also demonstrates the fruitfulness and utility of the Bayesian approach. It can explicate, and quite naturally, this episode of Dirac theory, which is otherwise explicable only in the vaguest sense.

Interestingly, the outcome of the story was that the experimental results were incorrect. A totally unexpected source of background electrons, due to plural scattering, precluded observation of the predicted effect. This was discovered during the early 1940s and the discrepancy between theory and experiment resolved. The experimental results on double scattering of electrons were both fallible and corrigible. Even a solution of the Duhem-Quine problem is uncertain. But, it can also be reasonable.

Department of Physics
University of Colorado

NOTES

1. E.R. Cohen and J.W. DuMond, "Our Knowledge of the Fundamental Constants of Physics and Chemistry in 1965," *Reviews of Modern Physics, 37* (1965), 537-94.
2. Particle Data Group, "Review of Particle Properties," *Reviews of Modern Physics, 48* (1976), S1-S245.
3. As M. Henrion and B. Fischoff ("Uncertainty Assessment in the Estimation of Physical Constants," forthcoming) suggest, this is, in part, due to the fact that scientists tend to underestimate their uncertainties. This is true, but it does not explain the entire effect. See Cohen and DuMond, note 1.
4. E. Fermi, "Attempt at a Theory of β-Rays," *Nuovo Cimento, 11* (1934), 1-21.
5. Fermi considered "allowed" decays, those for which the electron and neutrino wavefunctions could be considered constant over nuclear dimensions. "Forbidden" transitions also exist, and their rate is very much reduced. The shape of the decay spectra for such transitions differs from that of "allowed" transitions, a point we shall return to later.
6. B.W. Sargent, "The Maximum Energy of the β-Rays from Uranium X and other Bodies," *Proceedings of the Royal Society (London), A139* (1933), *659-73.*

7. E. Konopinski and G.E. Uhlenbeck, "On the Fermi Theory of Radioactivity," *Physical Review*, 48 (1935), 7-12.
8. C.D. Ellis and W.J. Henderson, "Artificial Radioactivity," *Proceedings of the Royal Society (London), A146* (1934), 206-16.
9. B.W. Sargent, "Energy distribution curves of the disintegration electrons," *Proceedings of the Cambridge Philosophical Society, 24* (1932), *538-53.*
10. H.A. Bethe and R.F. Bacher, "Nuclear Physics," *Reviews of Modern Physics, 8* (1936), 82-229. Quotation on p. 192.
11. E.J. Konopinski and G.E. Uhlenbeck, "On the Fermi Theory of β-Radioactivity," *Physical Review*, 60 (1941), 308-20.
12. J.L. Lawson, "The Beta-Ray Spectrum of Phosphorus, Sodium, and Cobalt," *Physical Review, 56* (1939), 131-6.
13. E. Konopinski, "Beta Decay," *Reviews of Modern Physics, 15* (1943), 209-45. Quotation on p. 218.
14. A. Franklin, *The Neglect of Experiment* (Cambridge University Press, 1986), Chapter 6.
15. I. Hacking, *Representing and Intervening* (Cambridge University Press, 1983).
16. A. Franklin and C. Howson, "Why Do Scientists Prefer to Vary Their Experiments," *Studies in History and Philosophy of Science, 15* (1984), 51-62.
17. This might be referred to as the Sherlock Holmes strategy. As Holmes remarked to Watson, "How often have I said to you that when you have eliminated the impossible, whatever remains, *however improbable*, must be the truth." [A. Conan Doyle, *The Sign of Four*, in *The Annotated Sherlock Holmes*, W.S. Baring-Gould, (ed.), (New York: Clarkson N. Potter, 1967), p. 638].
18. D. MacKenzie, "From Kwajalein to Armageddon? Testing and the Social Construction of Missile Accuracy," forthcoming.
19. This assumes, of course, that whiteness is not a defining characteristic of being a swan. Note the presence of black Australian swans at Leeds Castle.
20. See S. Harding (ed.), *Can Theories be Refuted*? (Dordrecht: D. Reidel, 1976).
21. For a detailed history of this episode see A. Franklin, "The Discovery and Nondiscovery of Parity Nonconservation," *Studies in History and Philosophy of Science, 10* (1979), 201-57.
22. C.S. Wu et al., "Experimental Test of Parity Nonconservation in Beta Decay," *Physical Review, 105* (1957), 1413-15.
23. A detailed history of this episode appears in A. Franklin, "The Discovery and Acceptance of CP Violation," *Historical Studies in the Physical Sciences, 13(2)* (1983), 207-38.

24. J. Christenson et al., "Evidence for the 2π Decay of the K_2 Meson," *Physical Review Letters, 13* (1964), 138-40.
25. J. Prentki, "CP Violation," Oxford International Conference on Elementary Particles, 1965, *Proceedings* (Oxford, 1966), 47-58.
26. Plausibility was not a requirement for offering such an alternative explanation. One of them involved a "shadow universe" which interacted with our universe only through the weak interactions. This alternative was not only testable, but was tested. See Franklin (note 23).
27. See Franklin (note 21).
28. There was another anomaly for Dirac theory at this time, involving the spectrum of hydrogen. This did not become important until the 1940s. For details see M. Morrison, "More on the Relationship between Technically Good and Conceptually Important Experiments: A Case Study," *British Journal for the Philosophy of Science, 37* (1980), 101-15.
29. There are other types of important experiments, but that is beyond the scope of this paper.
30. I am assuming here that $K\varepsilon$ is much greater than ε. The argument is still valid if $K\varepsilon$ is either approximately equal to or is much less than ε.

III

THE ROLE OF THEORETICAL CONCEPTIONS

CATHERINE CHEVALLEY

PHYSICAL REALITY AND CLOSED THEORIES IN WERNER HEISENBERG'S EARLY PAPERS

In his 1934 paper entitled "Transformations of the foundations of natural science"[1] Heisenberg stressed the possibility for physical theories to "exhibit a structure different from the structure of classical physics". This, he said, would happen in the one case when their objects would not be any more "things encountered in a direct and immediate experience"[2]: to leave this field ruled by classical physics would thus generate a break, a change, a shift in the very structure of the physical theory. The "mimetic" relationship between the objects of the theory and the objects of our world would be lost and would not ensure any more the truth value of the propositions of natural science.

No one who is familiar with Heisenberg's concern in the 1930's about physical "reality" and the foundations of objectivity[3] will wonder at what he used to call his "erkenntnistheoretische" turn of mind[4]. But what is the connection of these views with his early work in quantum mechanics? In other words, is it possible to relate the statement quoted above with Heisenberg's specific approach of the problems met by physicists in 1926 and 1927? Making that assumption, I will attempt to show how Heisenberg's analysis of classical objectivity as linked to "the concept of the thing" is the development of the two suggestions he made at the beginnings of quantum mechanics: that the new theory should essentially be concerned with the "kind of reality" of its objects, and that one had to break with the former conception of the structure of physical theories.

THE "KIND OF REALITY" OF THE ELECTRONS (1926)

The existence of internal contradictions is in 1926 the most striking feature of physical research in atomic theory. There had been contradictions between the formalism and the concepts and images traditionally used in the interpretation since Planck's formula for the black-body radiation, but things had worsened afterwards with Bohr's atomic model.[5] Such difficulties created an inconsistency in the premises of the interpretation: despite the

D. Batens and J. P. van Bendegem (eds.), Theory and Experiment, 159–176.

obvious failure of pictures, these premises still referred to pictorial themes such as particle and wave, continuity and discontinuity, and many physicists persisted in thinking that only a choice of one or the other could provide the thread out of the labyrinth. The year 1926, during which Schrödinger published his four papers on wave theory and Born his statistical interpretation, was therefore a year of endless struggles about the nature of the physical reality which atomic physics was discovering.[6] Which physical entity was the more loaded with "physical reality", was it the particle or the wave? How should quantum theory be presented, with the accent on discontinuities or with the point of view of waves being the only reality, as Schrödinger insisted when he urged physicists to discard "abstract concepts such as transition probabilities, energy levels and the like?[7] What could be the reality of the stationary states and of the so-called quantum "jumps"? These questions became the more crucial after a demonstration was given of the mathematical equivalence of Heisenberg's matrix mechanics and Schrödinger's wave theory (by Schrödinger himself, and then by Pauli and Dirac). Schrödinger's point of view was strongly put, as shows this statement of his: "it has been doubted whether the phenomena in the atom could be included in the space-time form of thought at all; from the philosophic point of view, I would regard *as a complete surrender* a definitive decision in this sense, since we know that the forms of thought do not actually change".[8] However to sustain this kantian attitude, which implied that a realist representation of the phenomena could be given, led him to meet with many technical difficulties. In a letter addressed to him on May 27, 1926, Lorentz pointed out that he certainly preferred wave mechanics to matrix mechanics because it had a "greater intuitive clarity", but that nevertheless it seemed to him impossible to understand how a wave packet pulled apart by the dispersion in the medium would represent correctly a "particle", more generally "things to which we want to ascribe a rather permanent individual existence".[9] Other objections were presented.[10] During the fall of 1926, Schrödinger went to Copenhagen to give lectures and had heated discussions with Bohr. As Bohr wrote to Fowler on October 26, the conversation "centered gradually on the problem of the physical reality of the postulates of atomic theory"[11] but had no issue. It was clear, for instance, that something as simple as the path of an electron through a Wilson chamber could not be explained by Schrödinger's wave mechanics, which was the only interpretation available using classical pictures.

In a paper published in November as "Quantenmechanik"[12], Heisenberg proposed to reject entirely the very premises of these discussions. He pointed out that they were based on the belief that one should obtain an ontological interpretation of either the wave or the particle, while nobody seemed to be concerned by the basic meaning of the "physical reality" referred to. He suggested that a concept useful in classical physics may not be so in the world of atomic phenomena. And finally he inferred that a necessary condition to find a way out of the internal contradictions at issue was to analyse the word "reality". Such is therefore the task of quantum mechanics:

> "Investigation of the kind of physical reality which is proper to electrons and atoms is precisely the subject of atomic physics and thus also of quantum mechanics (...) It is the essential problem of atomic physics and consequently also the content of all quantum-mechanical considerations"[13].

In order to understand fully this statement, it is useful to recall the course of the argument. The reason why we encounter such difficulties - Heisenberg says - is that there is a contradiction between our "customary intuition" [14] and "the recent experimental discoveries". By the first, Heisenberg means classical physics (that is Newtonian mechanics and Maxwell's electromagnetic theory), namely a view "in which space and matter are in the last analysis imagined as continuous and decomposable in principle into arbitrarily small parts".[15] In this continuous framework, events "happen" and solid objects "move". But this customary representation of the world has been contradicted by some "direct experimental evidence of atomistic ideas", that is the various discoveries about the existence of a "typical discontinuous element" in processes which take place in "very small spaces and times" (for instance the fluctuation phenomena: Brownian motion, light scattering, and the experiments on particle beams: cathode rays, α- and β-rays). In front of this contradiction, physicists ask: what is the latent frame which generates its opposite, is it continuity or discontinuity? They would like to take the obvious way out of this trap, which is to confide into naïve atomism, in other words "to ascribe to these basic building blocks of matter - that is finally the positive and negative electron - *the same type of reality* as we more or less ascribe to objects of the daily world which surrounds us".[16] Then the atomic phenomena would "move in the familiar continuous space-time world which corresponds to our intuitive perception". However, this idea "has proven wrong in the course of time", since the new discontinuities cannot in any way be explained with the help of classical concepts. Such are for instance the discrete stationary states in the atom or the discontinuities in the radiation phenomena. Or even, more simply, the observational knowledge that, though we see the track of an electron in a Wilson chamber, we find it reflected on a diffraction grating like a wave. And "the language of classical physics [is] no longer capable of expressing these two observations as effects of a single entity", as Heisenberg will put it later on.[17] One must thus admit that "the electrons and the atoms do not have that degree of direct reality as the objects of our daily experience".[18] Physics cannot any more make a living out of naïve atomism. The task of quantum mechanics will be to explain the discontinuous atomic phenomena with something else than a reference to solid objects.

To summarize, the contradiction is between "intuition" and experiment. Despite the new data, physicists keep asking classical questions, striving to embody physical reality into only one picture. This leads to byzantine discussions. Therefore one must conclude that the nature of the questions should be changed. Instead of asking what entity is the more real, one should ask: what does the word "physical reality" mean in the case of atomic phenomena? This reversal of the question will become afterwards an essential feature of Heisenberg's methodology: if questions do not look too good, to put them upside down and always to wonder about the meaning of the words.[19]

Having liberated the interpretation of atomic phenomena from the classical notion of physical reality, Heisenberg is now free to suggest that the tension between intuition and experiment will disappear if one reforms the intuition:

> "The program of quantum mechanics must be first of all to liberate itself from the intuitive pictures, and, in place of the previously used laws of classical kinematics and mechanics, to establish simple relations between experimentally given quantities".[20]

To this program he adds a warning: physicists should avoid "to transfer classical concepts and ideas to the problems of atomic structure". Such a program however is not carried out in the 1926 paper; but it is soon after in the indeterminacy paper.

THE BREAK WITH THE STRUCTURE OF CLASSICAL PHYSICS (1927)

Heisenberg published his formulation of the uncertainty principle in the spring of 1927. The paper, entitled "On the intuitive content of the quantum-theoretical kinematics and mechanics"[21], appears to be the direct continuation of Heisenberg's new approach of the problem in the midst of the 1926 debates over the ontological basis of quantum theory. Beyond its significance for the history of physics on which I shall not comment here, it achieves a break with the classical architecture of a physical theory in the two ways announced in 1926.

First of all, the basic feature of physical theory in the XIXth Century epistemological constructions, namely the possibility to reestablish in principle at any time a connection with the world of our perception, is dismissed as irrelevant to the context of atomic physics. Though the title of the paper is misleading at first sight, the expression "intuitive content" is given an entirely new meaning:

> "We believe that we understand a physical theory intuitively when we can qualitatively conceive of the experimental consequences of this theory in all simple cases and when simultaneously we have recognized that the application of the theory never contains internal contradictions".[22]

Therefore, given this new definition, a physical theory can be "intuitive" even if it contradicts "our space-time concepts". Intuitivity becomes mathematical consistency plus a control over the area of experiments to which the theory is relevant. In other words, it is mathematical consistency plus a set of rules for the experimental application of the theory.[23] The contradictions which were due to "the struggle of opinions concerning discontinuum and continuum theory, particles and waves"[24], will disappear as soon as one does not any more require from the theory that it should be interpretable in visualizable concepts. Heisenberg's paper ends with the conclusion that "since we can think about the experimental consequences of the theory qualitatively in all simple cases, quantum mechanics need no longer be regarded as non-intuitive and abstract"[25]: this last comment is an answer to Schrödinger, who had said that he "felt discouraged, not to say repelled, by the methods of transcendental algebra, which appeared very difficult to me, and by the lack of visualizability".[26] But the sharpness of Heisenberg's move away from classical "Anschaulichkeit" also provoked Lorentz to complain bitterly during the Solvay Conference: "pour moi - Lorentz said - un électron est un corpuscule qui, à un instant donné, se trouve en un point déterminé de l'espace, et si j'ai eu l'idée qu'à un moment suivant ce corpuscule se trouve ailleurs, je dois songer à sa trajectoire, qui est une ligne dans l'espace (...). Je voudrais conserver cet idéal d'autrefois, de décrire tout ce qui se passe dans le monde par des images nettes"[27] ...

The 1927 paper allowed physical theory to break with the compelling goal of having at least in principle some link with the world of common experience (or rather with what is considered to be such). It also established new features for the fundamental concepts of the theory. The first part of the paper sets up definitions of words like "position", "trajectory", "momentum" and "energy" without any direct reference to the corresponding definitions in classical physics; the meaning merges out of a specification of the relevant experiments. Quantum-mechanical concepts do not refer to any pictorial image, they refer to the measurement process which leads to consider the quantities that they symbolize. For instance, if one "wants to clarify what is to be understood by the word "position of an object" (e.g. of an electron), one must specify particular experiments by which one intends to measure the "position of the electron"; otherwise the word has no meaning"[28] There could be no understanding of the word "position" without a knowledge of the Compton effect and the indeterminacy that it exhibits.[29] More generally, Heisenberg says, all definitions in quantum mechanics have to include the consequences of the uncertainty principle exactly as in Relativity theory the word "simultaneous" cannot be given a definition otherwise than through experiments for which the speed of propagation of light is essential.

As for Einstein, this new way of defining concepts should not be misunderstood as part of an operationalist view on physics in general. Its function is essentially a negative one; it allows the theory to get rid of any side-ways involvement with classical interpretations of kinematics. It creates a tabula rasa for the basic understanding of phenomena as "simple" as motion. The point here is not to promote a positivistic conception of physical theory, but only to break with the classical imagery.[30] Similarly the principle of a "reduction to observable quantities" will be described as a strategy, not as an epistemology, in the "Rapport sur la mécanique des quanta" which Heisenberg and Born deliver at the fifth Solvay Conference in October 1927.[31] Only by recalling this strategical feature of Heisenberg's method can one understand his insistence that the foundations of objectivity should be thoroughly questioned: why should one need such a drastic methodology if not because the issue at stake is the faith in the universality of classical physics? And why does the structure of classical physics appear universal, if not because its formalism is embedded with a set of interpretative statements which have no necessity of their own? Quantum mechanics does not break with classical physics, but with a certain kind of pervasive means of its legitimation. Hence appears a fourth task to be performed: after focusing on the kind of reality of its objects, liberating itself from intuitive pictures and establishing simple relations between experimental quantities, the new theory must provide an explanation of how classical physics invented the concept of "objective reality" as an extension of its own presuppositions.

CLASSICAL OBJECTIVITY AND THE CONCEPT OF THE THING

On many occasions at the beginnings of the 1930's, Heisenberg alludes to the "concept of the thing" (Dingbegriff) in a way which appeared probably quite mysterious, though it is very clearly connected with Heidegger's own contemporary interest in the expression.[32] Heisenberg writes for instance in his paper "Prinzipielle Fragen der modernen Physik"[33] that "classical physics represents the *clearest expression of the concept of the thing* in that it attempts to make the description of the world as independent as possible of our subjective experience". The concept of the thing here is what encompasses the objects of our world as permanent substances bearing properties which can change and which are the determinations of its mode of existence.

Objectivity and the thing are thus linked to one another. What is objectivity? It appears to be best described as: a postulate required at the foundations of classical physics. In order to understand this, one must turn to the first extensive description of the features of classical physics written by Heisenberg in 1934, where it is interpreted as the development of one basic assertion about things in space and time:

> "Classical physics, which reached its conclusion some thirty years ago, was built on some fundamental presuppositions which appeared to be obvious starting points of all exact science, and seemed to require neither proof nor discussion: *physics dealt with the behavior of the thing is space and its change in time*".[34]

The kind of physics which met its full achievement and at the same time its limits in 1900 or 1905 – "some thirty years ago" – with Planck's introduction of the quantum of action or Einstein's special relativity had become the model for all exact knowledge about the world. However, Heisenberg says, this preeminence was derived from the illegitimate claim that physics was describing a state of affairs, the behavior of things in space and their change in time, and that its presuppositions were "obvious". There was a tacit assumption that some properties which were characteristics of a given set of experiences possessed universal validity:

> "Although this [the description of things] originally characterized *only* the experiences which formed the basis of physics, nevertheless some properties of the thing were inferred from these experiences and seemed to be determined at the same time. One was led to the tacit assumption that there existed an objective course of events in space and time, independent of observation: further, that space and time were categories of classification of all events, completely independent of each other, and thus represented an objective reality, which was the same to all men".[35]

As an extension of the concept of the thing, "objective reality" is essentially linked with classical causality. In classical physics, Heisenberg says, one starts from the observation that a thing always can be localized in a given point of space at a given time and that some regularities occur, which affect the properties of the thing without modifying its permanent substrate. Once such a generalization is made, the thing (Ding) is transformed into an object (Gegenstand) for physical theory: the immediately given has been represented in a universal proposition. To Heisenberg, this scheme is fully expressed into the formalism of classical physics. From the complete knowledge of what is a thing (that is of the initial state of a system), one will derive the knowledge of its future state at any time: this is what Heisenberg calls causality in 1927 and 1931.[36] Therefore, he writes that "in classical physics a problem was considered solved when a particular phenomenon had been proved to occur objectively in space and time and had been shown to obey the general rules of classical physics as formulated by differential equations".[37] In other words: our "common experience" is that things in ordinary space and time obey regularities; this leads to a causal representation of their behaviour in the formalism of physics; ultimately, in order to insure the universal validity of physics, the concept of an independent reality identical for all men is formed.

As a consequence of this objectivation of the thing, physics will require visualization of its processes. Then, transcending this methodological level, it will claim to provide an image of the world, a general "Weltanschauung", based on the belief that mathematical physics has the ability of telling the truth about what is "real". Thus interfering into other spheres of intellectual life, exact science "threatened its own status"[38]: questions and problems which would be unintelligible if stated following the principles of classical physics could simply not be formulated.

PHYSICAL THEORIES AS CLOSED THEORIES

These features of classical physics are the work of XIXth Century. XIXth Century originated, Heisenberg says, the irrational belief that "one could gain an understanding of the whole world from a small part of it".[39] However, by its very existence, quantum mechanics plays the part of a "factum rationis": it gives evidence that one may have a physical theory without having to look for the support of an ontology of things, that is of a doctrine of how things exist. Quantum mechanics "makes phenomena 'objective' without the demand for a 'thing in itself' capable of description in terms of everyday experience".[40] Moreover there is no privileged connection whatever between the things and the objects of physics; some physical theories deal with the objects which we see every day, others do not. A physical theory can have objects which are not things, that is not interpretable as visualizable permanent substances. Conversely the things may very well have an existence of their own and a truth of their own: for instance, Heisenberg says, the goal of Goethe's "bitter struggle" against Newton was to recall that reality could overstep the "cold and abstract concepts" of classical mechanics.[41] The conclusion would be here that one must avoid to fall into an epistemological illusion: the phenomenon as the objectivation of the things. The path of the electron in a Wilson Chamber is for that matter a phenomenon of a new kind.[42]

Classical physics is thus essentially suffering from its pretense of universality. Once deprived of it, it appears as a well-defined conceptual system, which Heisenberg describes in 1935 as follows: "a system of mathematically concise axioms whose physical content is fixed by the choice of the words used in them; hence they determine unequivocally the application of these systems of axioms to nature".[43] For Heisenberg, this became the correct definition for any physical theory: to be mathematically consistent in a given well-defined area of experiments; to have a vocabulary with non-ambiguous definitions of the words (at least for the observational terms); and to have a correspondence set of rules for the application of the formalism to the phenomena belonging to the area. Heisenberg here clearly anticipates the conception of a physical theory as a partially interpreted formalism as one finds it in later developments of logical empiricism, even though his philosophy of science is basically different.

Since logical consistency is obviously an essential feature for physical theories, it is not surprising that Heisenberg chooses the expression of "closed theories" ("abgeschlossene Theorie") to qualify them. A "closed" theory would be a theory in which no equivocity or ambiguïty would hamper the application of the propositions of the theory to the phenomena. He uses this expression as soon as 1926[44]; in a lecture given in Vienna in 1930, he stresses the idea that quantum mechanics is "epistemologically a closed theory"[45]; in the subsequent papers, he repeatedly stresses the fact that "modern physics has shown that the structure of classical physics, as that of modern physics, is closed in itself"[46] and finally he devotes two entire papers to this topic in 1948 and 1972, together with the sixth chapter of *Physics and Philosophy*.[47]

However, being a central issue in Heisenberg's philosophy of natural science does not prevent the notion of "closed theories" to generate as many problems as it solves, all of them in connection with the question of what is axiomatization for physical theories. In his interview with T.S. Kuhn, Heisenberg recalls that, due to the influence of Hilbert, the idea gradually developed in the minds of many physicists "that we may be forced to describe nature by means of an axiomatic system which was thoroughly different from the old classical physics". But the reference problem cannot be stated in the same way in a physical theory and in a mathematical theory. Heisenberg has attempted, mostly in later papers, to give a formulation of what could be a possible legitimation for an axiomatic physical theory which lacks any support in perception, as it is the case in quantum mechanics. At the very same time, in September 1936, Jean Cavaillès expressed similar worries in a letter to Albert Lautman: "pour le fond, je crois que nous sommes de plus en plus d'accord: extérieurement rien ne distingue un système axiomatisé mathématique ou physique. Reste la jonction avec le réel: en mathématique, l'expérience est décrite exhaustivement par les axiomes; la pensée n'est distincte du réel que par l'arbitraire qu'elle reconnaît dans la réunion des axiomes. En physique, on revient au sens classique d'expérience-approximation: ce n'est pas très satisfaisant".[48]

QUANTUM MECHANICS WITHOUT CLASSICAL REALITY

In the decade which follows his first contributions to physics, Heisenberg is thus striving to give full clarity to the fact that, as N.R. Hanson put it, "nature refused to live up to XIXth century expectations".[49] His analysis is rooted in his work of 1926 and

1927; it also sets up the basis of the further trend of his philosophical thought.

Classical physics was supposed to be the description of a world of things, which was given the name of "objective reality". These things had a definite way of behaving in space and time, and this could be accounted for either in ordinary language with the help of appropriate images, or in the language of mathematics. Therefore classical physics was composed of mathematical physics with the addition of an ontology which played the part of asserting the truthvalue of its statements. From this, there sprung during the XIXth century a general "view of the world", a Weltanschauung supposedly grounded in the exact sciences. But the emergence of quantum mechanics destroyed the confidence into such an homogeneous relation between things and the objects of physics. It showed that classical physics was comprised of a set of closed theories, ruling definite areas of the field of human experience as well as art, literature of religion did also produce "axiomatized" representations and styles.[50]

On the other hand, quantum mechanics, away from things, dealt with phenomena which could not be understood with the help of the "ordinary space-time concepts" and had no intuitive foundation in the classical sense. No visual models could help to explain the "atom of modern physics", since it was nothing more "than a partial differential equation in a multidimensional abstract space".[51] The physical phenomena could not be any more conceived as permanent bearers of properties: "quantum mechanics is in no way concerned with the objective determination of space-time phenomena"[52], but it deals with wave functions, distributions of probabilities and perturbations. The concept of an objective reality independent of observations and composed of individual elements has to be put aside: "Nature, through the medium of modern physics, has reminded us very clearly that we should never hope for such a firm basis for the comprehension of the *whole field of perceptible things*".[53] In other words one has "to forego visualization and the process of objectivation" ("Anschaulichkeit und Objektivierbarkeit")[54], but it does not mean anything like subjectivism. Heisenberg's view is to differentiate "levels of reality" and "concepts of reality". All levels of reality should be defined in connection with the description of the observational processes involved; for instance "laws of nature" must include constants like c and h to serve as standards of measurements.[55] Then the different and sometimes overlapping sectors of reality could be classified by paying attention to "the type of question we ask of nature and the amount of interference which we allow to observation".[56] The so-called division of the world between subjective and objective is, says Heisenberg, far too simple. But one cannot replace it without differentiating also the concepts of reality.

In later papers[57], Heisenberg will suggest to pick up the kantian distinction between "Realität" and "Wirklichkeit". A physical theory always deals with "actual reality" (Wirklichkeit), that is effective existence, that which we can have an external perception of; and so does quantum mechanics: "the actual plays the same decisive part in quantum theory as it does in classical physics. The Copenhagen interpretation is indeed based upon the existence of processes which can be simply described in terms of space and time, that is in terms of classical concepts, and which thus compose our "reality" in the proper sense".[58] However, in atomic physics, though we perceive only actual processes, the mathematical formalism describes a set of possibilitьes which constitutes another level of reality (Relität), parts of which are brought to "actuality" through the interaction. Such a distinction should abolish the belief that physics is or can be a complete and true translation of what happens in nature: "the ontology [of classical physics] rested upon the illusion that the kind of existence, the direct "actuality" of the world around us can be extrapolated in the atomic range".[59] An expected consequence of this situation is that it becomes meaningless to derive any "Weltanschaung" or general view of the world from physics itself: "physics cannot be any more the starting point from which all questions of Weltanschauung are attached".[60] Another consequence, much more difficult to cope with, is that it becomes impossible to speak truthfully about atomic phenomena in the ordinary language: ordinary language is stuck with things, since language was formed in order to provide help in our encounters with them. Therefore there is an oblique, ambiguous, vague way of speaking about the objects of quantum theory, a way of speaking which transforms the classical concepts into pictures and metaphors.[61] The lack of any "original rectitude of the names" leaves the quantum theorist pretty much in the position of Cratylus in Aristotle's *Metaphysics*: Cratylus wants to make clear that nothing can be said truthfully about the world because no thing is permanent, and therefore he remains silent and only moves his finger around to show the creation and annihilation of all substances. Fortunately, one may say, the quantum theorist can speak because he has complementarity.

Going back to Heisenberg's statement that physical theories will exhibit a different structure if there is a change in the kind of reality of their objects, it is clear that the notion of "closed theories" is but the development of the 1926–27 discussions of the problems met by atomic theory. Starting from a denial of the belief that one could retain the classical notion of physical reality in the world of atomic phenomena, Heisenberg contributed in an essential manner to the formulation of quantum mechanics as a physical theory "intuitive and complete" (by giving entirely different meanings to those characteristics); then his view of classical physics as the development of the concept of the thing allowed him to link the new physics to a criticism of modern ontology. All physical theories - including classical physics - ultimately appear to be closed theories, bearing their own ontologies, with no claim to "more" universality for one or the other, nor for physics versus art, religion or philosophy. Quantum mechanics has thus the effect of unveiling the functional architecture of classical physics - a mathematically consistent set of propositions - hidden under the apparent structure of a "true" knowledge about nature in general. These complex assertions are, in some way, meaningless for the development of physics itself; moreover, they have often been reduced to such and such current trend of philosophical thinking. It was enough here to show the consistency of Heisenberg's views with his work as a physicist. A further analysis, however, would have to interpret these views in respect of the history of philosophy.

CNRS, Paris
Boston University, Boston

NOTES

1. Heisenberg (1934). Quoted from Heisenberg (1945).
2. Heisenberg (1945), 17. All citations will be from the German texts; most translations are mine.
3. See Heisenberg (1931), (1933), (1934a,b), (1937), (1941), (1942), for samples of Heisenberg's extensive historical and philosophical evaluations of classical physics, Greek philosophy, German romanticism, or Goethe's theory of colour.
4. See P. Heelan (1965) and (1975)., J. Hendry (1984), H. Hörz (1968).
5. See J.L. Heilbron (1969) and A.I. Miller (1984).
6. For an account of the intensity of these struggles, see M. Beller (1983).
7. Schrödinger (1926a). Quoted from G. Ludwig (1968).
8. Schrödinger (1926a). Quoted from G. Ludwig (1968), 119-120.
9. Quoted from M.J. Klein (1967), 43.
10. See M. Jammer (1974), 24-33.
11. N. Bohr (1985), VI, 423.

12. Heisenberg (1926b).
13. Heisenberg (1926b), 989.
14. The German expression is "gewönliche Anschauung". See essentially A.I. Miller (1984), 125-183.
15. Heisenberg (1926b), 989.
16. Ibid., 989. All quotations are from the introduction of the paper.
17. Heisenberg,(1936), 100.
18. Heisenberg (1926b), 989.
19. See for instance Heisenberg (1958), chapter 2, and Heisenberg (1975). Also, about the meaning of the word "division" applied to particles in high-energy physics, see Heisenberg (1976).
20. Heisenberg (1926b), 990.
21. Heisenberg (1927a).
22. Heisenberg (1927a), 172.
23. See also Heisenberg (1927b), for similar assertions.
24. Heisenberg (1927a), 172.
25. Heisenberg (1927a), 196.
26. Quoted from G. Ludwig (1968), 128.
27. See *Electrons et Photons* (1928), 248.
28. Heisenberg (1927a), 174.
29. As a matter of fact, the description given by Heisenberg for the Compton effect generated a disagreement with Bohr about the meaning of the indeterminacy principle versus classical physics. But this does not affect the methodological trend which I consider here.
30. This assertion will be argued more at length in a subsequent paper.
31. See Heisenberg (1927b), 143-144: "La mécanique des quanta essaie d'introduire les nouvelles notions par une analyse précise de ce qui est "essentiellement observable". Cela ne revient pas à établir le principe qu'il est possible et même nécessaire de faire une séparation nette entre ce qui est "observable" et ce qui est "inobservable". Dès qu'un système de concepts est donné, on peut conclure des observations à d'autres faits qui à proprement parler ne sont pas directement observables, et la limite entre ce qui est observable et ce qui ne l'est pas devient tout à fait indéterminée."
32. Heidegger's *Die Frage nach dem Ding* was initially a course of lectures held at the University of Freiburg in 1936. Heidegger and Heisenberg first met in the fall of 1935. I thank D. Cassidy and H. Rechenberg for making enquiries about the possibility of an existing correspondence, with results which have been negative up to now. A plausible conjecture is that Heisenberg found in Heidegger's interpretation of modern onto-

logy the most powerful expression of his own views about the signification of quantum mechanics, at least at this time. This connection will be explored more thoroughly in another paper. My main goal here is only to establish a link between Heisenberg's work in 1926–27 and his general interpretation of quantum theory as developed in the 1930's.

33. Heisenberg (1936). Quoted from Heisenberg (1945), 42. The English translation in *Philosophic problems of nuclear science* gives "concept of matter" for "Dingbegriff".
34. Heisenberg (1934). Quoted from Heisenberg (1945), 7.
35. Heisenberg, ibid.
36. See Heisenberg (1927a), 197: "Aber an der scharfen Formulierung des Kausalgesetzes: 'wenn wir die Gegenwart genau kennen, können wir die Zukunft berechnen', ist nicht der Nachsatz, sondern die Voraussetzung falsch". See also Heisenberg (1931), quoted in M. Jammer (1974), 76, as: "if at a given time all data are known for a given system then it is possible to predict unambiguously the physical behavior of the system also for the future".
37. Heisenberg (1934b).
38. Heisenberg (1934a). Quoted from Heisenberg (1945), 19.
39. Heisenberg (1934). Quoted from Heisenberg (1945), 19
40. Heisenberg (1942). Quoted from Heisenberg (1945), 93. The 'thing in itself' is a reference to Kantian philosophy; but Kantian philosophy, or rather its rigid interpretation, has been serving as a tool for the "dogmatization of scientific concepts": see for this Heisenberg (1934).
41. Heisenberg (1941). Quoted from Heisenberg (1945), 58.
42. Bohr stresses similarly the necessity to redefine the word "phenomenon" by associating its meaning to the specification of the corresponding experimental device rather than to any empirical manifestation. See for instance "The unity of knowledge", in *Atomic Physics and human knowledge*, New York, J. Wiley, 1958; and E. Scheibe (1973), 20-23.
43. Heisenberg (1936). Quoted from Heisenberg (1945), 38.
44. Heisenberg (1926a), (1926b).
45. See M. Jammer (1974), 77-78.
46. See for instance Heisenberg (1934), in Heisenberg (1945), 20. Ibid: "the changes in the foundations of science, forced upon us by Nature (...) leave classical physics untouched, but they show that the scientific systems like classical mechanics or other parts of classical physics must always be closed in order to be correct".
47. See Heisenberg (1948), (1972) and (1955a).
48. Quoted from the Archives Suzanne Lautman, with permission.

49. N.R. Hanson (1959), 2.
50. The notion of "style" applied to physics is especially developed in Heisenberg (1955a), chapter 6.
51. Heisenberg (1933). Quoted form Heisenberg (1945), 36.
52. Heisenberg (1934b). Quoted here from Heisenberg (1965), 296.
53. Heisenberg (1934a). Quoted from Heisenberg (1945), 22.
54. Heisenberg (1934b). Quoted from Heisenberg (1965), 301.
55. Heisenberg (1958b). Quoted from Heisenberg (1974), 12.
56. Heisenberg (1941). Quoted from Heisenberg (1945), 74.
57. See especially Heisenberg (1955).
58. Heisenberg (1955), 28.
59. Ibid.
60. Heisenberg (1934a). Quoted from Heisenberg (1945), 22. Here again the connection with Heidegger's rejection of philosophy as a "Weltanschauung" should be stressed.
61. See T. Bergstein (1972) and C. Chevalley (1986).

REFERENCES

The list of references below is restricted to those mentioned in the footnotes. I have used the 1945 issue of Heisenberg's *Wandlungen in den Grundlagen der Naturwissenschaft* which is composed of (1933), (1934), (1936), (1937), (1941) et (1942). For a complete list of his publications, one can see *Werner Heisenberg, Gesammelte Werke - Collected Works*, eds. W. Blum, H.P. Dürr and H. Rechenberg, Berlin-New York, Springer-Verlag, 1984-, A. Original scientific papers. B. Scientific review papers, talks and books. And ibid. , Münich, R. Piper: C. Philosophical and popular writings. See also *Werner Heisenberg: a bibliography of his writings*, eds. D. Cassidy and M. Baker, Berkeley, Office for the history of science and technology, University of California, 1984.

Mara BELLER, "Matrix theory before Schrödinger. Philosophy, problems, consequences", in *Isis*, vol. 74 (1983), 469-491.

T. BERGSTEIN, *Quantum theory and ordinary language*, London, Macmillan (1972).

Niels BOHR, *Collected Works. Vol. VI: Foundations of Quantum Physics I (1926-1932)*, eds. E. Rüdinger and J. Kalckar, Amsterdam-New York-Oxford-Tokyo, North-Holland Publishing Company (1985).

Catherine CHEVALLEY, "Complémentarité et langage dans l'interprétation de Copenhague", in *Revue d'histoire des sciences*, t. XXXVIII-3/4 (1985), 251-292.

Norwood R. HANSON "The Copenhagen interpretation of quantum theory", in *American Journal of Physics*, 27, I (1959), 1-15. Reprinted in Stephen TOULMIN ed., *Physical Reality. Philosophical essays on XXth century Physics,* New York, Harper & Row (1970).

Patrick A. HEELAN, *Quantum Mechanics and Objectivity. A study of the physical philosophy of Werner Heisenberg,* La Hague, Martinus Nijhoff (1965)

"Heisenberg and radical theoretical change" in Zeitschrift für allgemeine Wissenschafttheorie, 6 (1975), 113-138.

John L. HEILBRON, "The Genesis of the Bohr Atom", in *Historical Studies in the Physical sciences,* 1 (1969) 211-290 (with Thomas S. Kuhn).

Werner HEISENBERG, 1926a "Mehrkörperproblem und Resonanz in der Quantenmechanik", in *Zeitschrift für Physik,* 38 (1926), 411-426.

1926b "Quantenmechanik", in *Die Naturwissenschaften,* 14 (1926), 899-994. English translation in Arthur I. Miller ed. *Source Book in relativity and Quantum Field Theory,* Harvard University Press (to be published in 1988).

1927a "Ueber den anschaulichen inhalt der quantentheoretischen Kinematik und Mechanik", in *Zeitschrift für Physik,* 43 (1927) 172-198. English translation in J.A Wheeler and W.H. Zurek eds. *Quantum Theory and Measurement,* Princeton University Press (1983), 62-84. Also in A.I. Miller, op. cit. above.

1927b "La Mécanique des Quanta" (with Max BORN), in *Electrons et Photons, Rapports et discussions du cinquième conseil Solvay,* Paris, Gauthier-Villars (1928).

1931 "Die Rolle der Unbestimmtheitsrelationen in der modernen Physik", in *Monatsheft für Mathematik und Physik,* Leipzig, 38 (1931), 365-372.

1933 "Zur Geschichte der physikalischen Naturerklärung, *in Physikalische Berichte* 85 (1933), 29-40, Reprinted in (1945).

1934a "Wandlungen der Grundlagen der exakten Naturwissenschaft in jüngster Zeit", in *Die Naturwissenschaften,* 22 (1934), 660-675. Reprinted in (1945).

1934b "Die Entwicklung der Quantenmechanik", in W. Heisenberg, E. Schrödinger, P.A.M. Dirac, *Die moderne Atomtheorie. Die bei der Entgegennahme des Nobelpreises 1933 in Stockholm gehaltenen Vorträge,* Leipzig, S. Hirzel, (1934), 1-18.

1936 "Prinzipielle Fragen der modernen Physik", in *Unsere Welt* 28 (1936) 97-102. Reprinted in (1945).

1937 "Gedanken der antiken Naturphilosophie in der modernen Physik", in *Die Antike. Zeitschrift für Kunst und Kultur des klassischen Altertums,* 13 (1937), 118-124. Reprinted in (1945).

1941 "Goethesche und Newtonsche Farbenlehre im Lichter der modernen Physik" in *Geist der Zeit* 19 (1941), 261-275. Reprinted in (1945).

1942 "Die Einheit des naturwissenschaftlichen Weltbildes", in *Leipziger Universitätsreden*, 8. Reprinted in (1945).
1945 *Wandlungen in den Grundlagen der Naturwissenschaft*, Leipzig, S. Hirzel. (same as the 1942 edition, but easier to find).
1948 "Der Begriff 'Abgeschlossene Theorie' in der modernen Naturwissenschaft", in *Dialectica*, 2 (1948), 331-336.
1952 *Philosophic problems of nuclear science*, New York, Pantheon; London, Faber & Faber. English translation of (1945).
1955 "The development of the interpretation of quantum theory", in *Niels Bohr and the development of physics. Essays dedicated to Niels Bohr on the occasion of his 70th birthday*, W. Pauli, L. Rosenfeld, V. Weisskopf eds. London, Pergamon, New York, McGraw-Hill.
1958a *Physics and Philosophy. The revolution in modern science*, New York, Harper & Row.
1958b "Die Plancksche Entdeckung und die philosophischen Grundfragen der Atomlehre", in *Max Planck-Gesellschaft, Jahrbuch*, 26-52, reprinted in (1974).
(1965) *Physics 1922-1941*, Nobel Lectures published for the Nobel Foundation, Amsterdam-London-New York, Elsevier.
1972 "Die Richtigheitskriterien der abgeschlossenen Theorien in der Physik", in *Einheit und Vielheit, Festschrift für C.F. von Weizsäcker zum 70 Geburtstag*. E. Scheibe and G. Sussmann eds., Göttingen, Vandenhoeck & Ruprecht, 140-145. English translation in W. Heisenberg, *Tradition in Science*, New York, The Seabury Press (1983), 123-129.
1974 *Across the frontiers*, New York, Harper & Row.
1975 "The Development of concepts in the history of quantum theory", in *American Journal of Physics*, 43 (1975), 389-394.
1976 "The nature of elementary particles", in *Physics Today*, 29, (1976) 32-39.

John HENDRY, *The Creation of Quantum Mechanics and the Bohr-Pauli Dialogue*, Dordrecht, Reidel, 1984.

Herbert HORZ, *Werner Heisenberg und die Philosophie*, Berlin, VEB Deutscher Verlag der Wissenschaften (1968).

Max JAMMER, *The Philosophy of Quantum Mechanics*, New York, John Wiley & Sons (1974).

Martin J. KLEIN, *Letters on Wave Mechanics. Schrödinger, Planck, Einstein, Lorentz*, K. Przibram ed. New York, Philosophical Library (1967). Introduction and translation by M.J. Klein.

Gunther LUDWIG, *Wave Mechanics*, New York, Pergamon (1968).

Arthur I. MILLER, *Imagery in scientific thought. Creating XX-th Century Physics*, Boston, Birkhauser (1984).

Erhard SCHEIBE, *The logical analysis of quantum mechanics*, Oxford, New York, Pergamon Press (1973).

GERARD SIMON

EXPERIMENT AND THEORY IN PTOLEMY'S *OPTICS*

Using examples from Ptolemy I intend today to interrupt a too exclusive exchange between experiment and theory implied by the very title of this colloquium. It seems to me necessary to examine matters further down, down to the thought objects underlying the theories, without overlooking the arrangements of knowledge and the cultural context (almost in an ethnological meaning) surrounding the scientist at work.

I shall deal with the case of an unrecognized experimental discipline, the optics of antiquity, and more particularly Claude Ptolemy's *Optics*.[1] In this very country, especially in Louvain, Albert Lejeune conducted a lot of valuable surveys about it, which have unfortunately remained too confidential.[2] Thirty to forty years later his findings are worth recalling, and some of his conclusions worth discussing which in my opinion modernize Ptolemy's thought objects unduly.[3]

In his *Optics* Ptolemy has used experimentation as means of discovery and proof in two cases:

- In the study of binocular vision (Books II and III). I shall deal mainly with this point, which has so far remained practically unknown.
- In the study of reflection and refraction, in order to establish equality between angles of incidence and reflection of visual rays (Book III); and to work out a numeric table (from zero to ninety degrees) of the refraction of these rays from air to water, from air to glass, from water to glass (Book V). I shall only recall this study, but I shall carefully consider to what objects these experiments are indeed related.

I. EXPERIMENTS ON BINOCULAR VISION

1. *The problem in antiquity and its significance*

We have first to understand that the problem of binocular vision is most important in the optics of antiquity, because ancient geometers follow the visual ray theory. Among the conceptions

D. Batens and J. P. van Bendegem (eds.), Theory and Experiment, 177–188.

of vision of their time, it is the only theory to establish a one-to-one correspondence between one element of visible and one element of the view. This correspondence is the basis of geometrical analysis of vision. Let us recall of what this hypothesis was.

We see when one of our visual rays, issued form the eye, goes and touches one point of the object; together the visual rays form a visual cone, the vertex of which is in the eye and the generative line follows the circle of the pupil. The direction of the point which is actually looked at is provided by the axis of the cone.

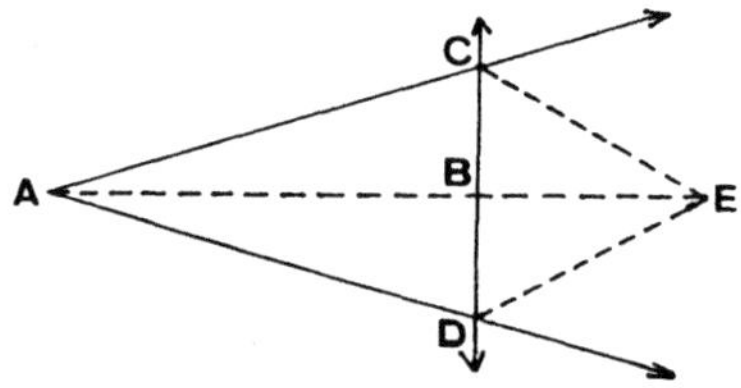

fig. 1.

As the ray *comes* from the eye and sets a direction of sight, any convergence in this ancient science of optics is disastrous for vision. Let us name for instance the eye A, ABE the axis of the visual cone, AC and AD two visuals rays, and E the object point (fig. 1.). If AC and AD are convergent to E under the effect of something like a lens, E will be seen simultaneously in the directions AC, AB and AD. What for a modern is the key to the formation of a sharp image by a lens or a concave mirror, or by the crystalline lens itself, is in antiquity an insufferable theoretical situation.

What is true for one eye is, of course, true for both. Why don't we perceive with both eyes the object in two different directions? That is a critical question on which the intellectual coherence of the discipline depends. Ptolemy's answer is the following: "Nature has instituted binocular vision in order to make it more powerful, and so that it works in a specific and orderly way. We are endowed with the ability to turn our eyes, as soon as we open them, towards various places, with a wonderful speed, until their axes hit together the middle of the body to be seen, and only one single base - common to the two visual cones - is formed by their contact with the object, made up of all the homologous rays" (Book II, 28). These homologous rays are those which have "a similar and [*aequalis*] - i.e. equal or equivalent - position relative to their axes": namely, each being situated on the same side of its respective axis, either on the left or on the right, and forming with it the proper angle.[4]

In Book II, in which he deals with direct vision, Ptolemy experimentally establishes the correspondence between the visual axes, the necessity of looking at convergence point in order to have a simple vision, and, should it be otherwise, the duplication of the vision. In Book III, which deals with reflection, he broaches the topic once more and experimentally establishes those rules according to which the fusion of the two visual cones occurs. I only mention here as an example the experiences of Book II.

2. *The experiments in Book II*

a. *First series of experiments*

* *Equipment*

Ptolemy uses for his purpose two cylindrical sighting marks, one white and the other black, both sliding on a small ruler; this kind of device is still in use nowadays for studying the same phenomenon. It was drawn from existing astronomical instruments. Both ruler and sighting marks are set perpendicular to the frontal plane, at the base of the nose, in the place that modern scientists call the cyclopean point.

* *First experiment*

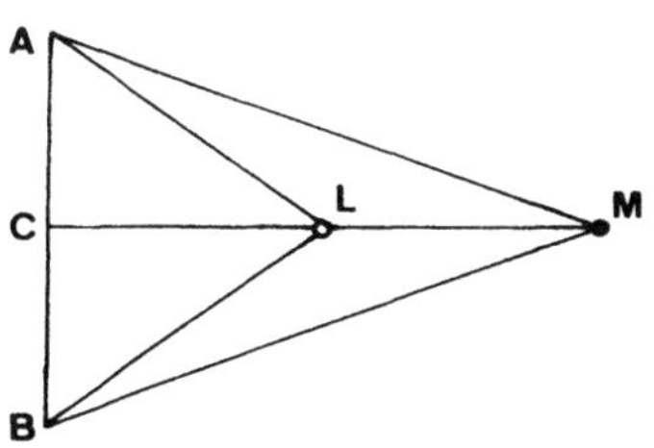

fig. 2

Key to fig. 2:

A = left eye
B = right eye
C = cyclopean point
L = white sighting mark
M = black sighting mark

1) Both eyes are fixed on L, white. M can be seen in two places, on the left and on the right of L:

Interpretation: the black cylinder on the left of L is M seen by A (M_A) and the right one, M seen by B (M_B). L is seen by both A and B (L_{A+B}).
2) Both eyes are fixed on M, black. L can be seen in two places on the left and on the right of M:

3) One of the eyes is closed (to check the interpretation).

If, when L is fixed, the left eye A is closed, M_A can no longer be seen on the left of L; and if the right eye is closed, M_B can no longer be seen on the right of L. We are therefore confronted with what is technically called *homonymous* diplopia: AM and BM rays are not homologous; in the left eye A, AM is situated on the left of the visual axis, and in the right eye B, BM is situated on the right of the axis.

If, on the contrary, when M is fixed, the left eye A is closed, L_A can no longer be seen on the right of M; and the right eye B, L_B on the left of M: we have the traditional case of *crossed* diplopia. AL and BL are not homologous, the former is situated on the right of the left visual axis, the latter on the left of the right visual axis.

Established as they are by experiment, Ptolemy's conclusions are anything but obvious: the importance of visual axes, and their usual homology; the duplication of the object looked at when positioned too far ahead or too far behind the fixation point; the classic distinction between homonymous and crossed diplopia.

A second experiment corroborates these conclusions.

* *Second experiment*

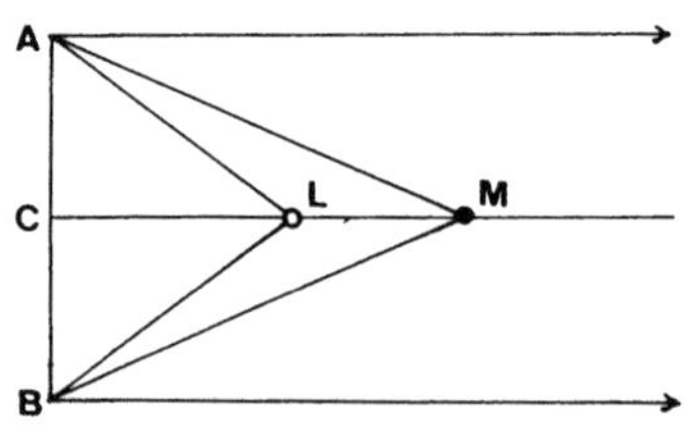

fig. 3

With the same equipment, both eyes are fixated on the horizon, on the infinity. What follows is a complete duplication:

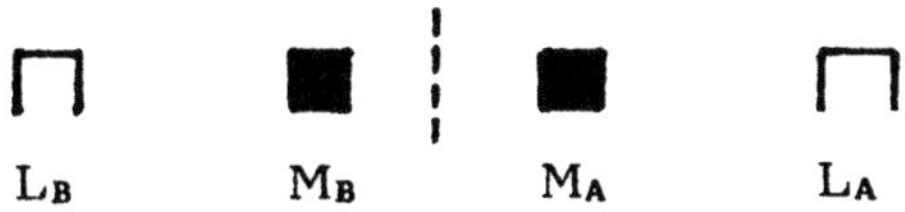

When the eye A is closed, the right cylindrical sighting marks disappear; and when B is closed, the left ones do.

This is the confirmation of the crossed diplopia interpretation. If there is no appropriate fixation of the visual axes, twofold vision ensues.

It must be noticed that these observations are not spontaneous; they can be done only be experiments which have to be *conceived*, which involves:

- building a device and appropriate equipment,
- following an explicit and constant procedure aiming to check the interpretation: the experimenter fixes his eyes on one point, closes one eye, then the other, without at any moment changing his point of fixation,
- using a gradual experimental approach so as to establish the facts.

This gradual approach is clearly noticeable in the second set of experiments.

b. *Second series of experiments*

The equipment is the same; but this time, the ruler is parallel to the frontal plane; and the observer's eyes are fixated on the horizon up to infinity. Three cases arise:

* *Third experiment*[5]

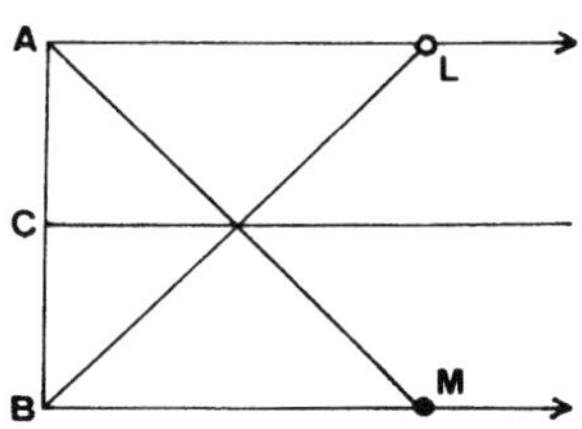

fig. 4

The distance LM is equal to the distance between the two pupils AB.

One can see L_B white, on the left, M_A black, on the right, and between them a third grey cylindrical sighting mark – a clear proof of the fusion of the two visual axes.

Eye A is hidden: M_A disappears, plus black component of the central figure.
Eye B is hidden: L_B disappears, plus white component of the central figure.

This proves that the two axes are homologous (there is a fusion of what they see) and that AM and BL are not homologous.

* *Fourth experiment*

The distance LM is superior to the interpupillar gap.

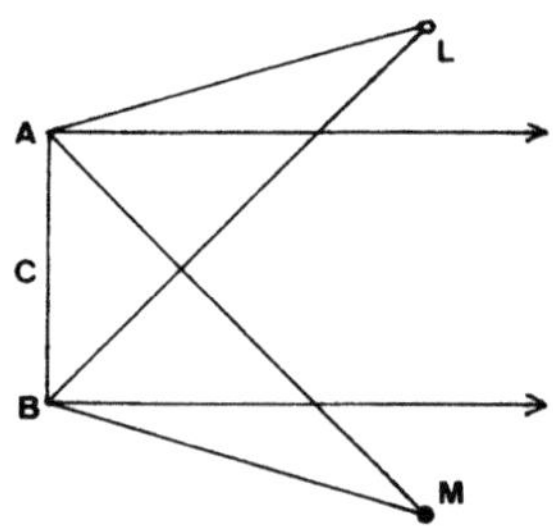

fig. 5

One can see two white cylinders on the left, and two black ones on the right:

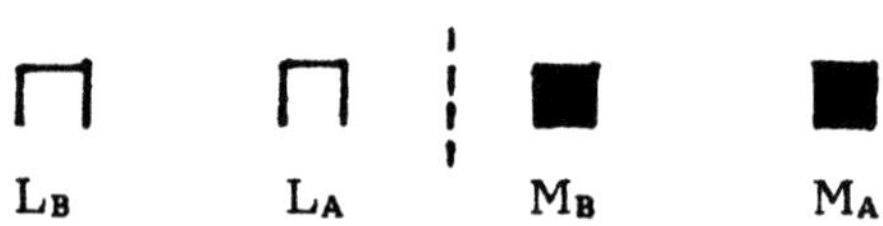

Eye A is hidden: L_A and M_A disappear.
Eye B is hidden: L_B and M_B disappear.
This corroborates homonymous diplopia.

* *Fifth experiment*

The distance LM is inferior to the interpupillar gap.

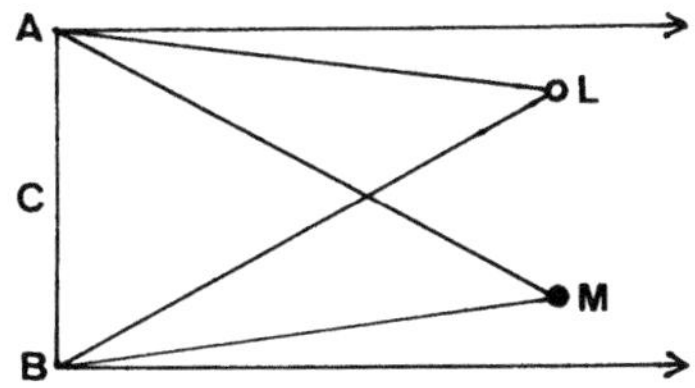

fig. 6

One can see alternate white and black cylinders:

Eye A is hidden: cylinders on the right disappear.
Eye B is hidden: cylinders on the left disappear.
This corroborates crossed diplopia.

The last two experiments show that when visual rays are not situated on the same side of their respective axes, both on the left or on the right, they are not homologous.

In Book III, a set of other even more ingenious experiments make it possible for Ptolemy to put forward a law of correspondence of visual cones – of a fusion of two cones into one. Let us merely note that the "horopter" (locus of all points seen simple at the same time for a given point of fixation) which he obtains is a fair estimate: the frontal straight line drawn at the point of fixation. Modern scientists, for the small distances, describe it rather as a curved zone comprising the fixation point at its summit, the concavity of which is turned towards the observer: the error does not exceed one or two centimeters from a distance of fifty to sixty centimeters and for a side distance in which observation remains clear.

3. *Experiments, thought objects and knowledge fields*

The explicit and intentional nature of the experiment is therefore evident. The development of an appropriate equipment, the working out of an unvaried procedure and the gradual approach used in the cases studied do appear clearly. However, Ptolemy's research cannot lead to the same notions and the same progress as ours.
a. First, the fusion of the visual cones is not a process similar

to the fusion of our two retinal images. The homologous rays fall on the *same* point of the object, which can be seen more clearly (as if it were doubly lit); but the base *common* to the two visual cones is strictly identical. In other words, it is not possible to draw an impression of *relief* out of the difference of retinal images, due to their respective points of view. At most each eye can see an enlarged part on the side of the common visual base. The solution to the sensation of relief is different from ours: as each visual ray provides for the consciousness of its own length, it is as if the eye were performing a broad palpation from a distance. So the visual cone is a kind of sensorial organ in itself, springing out of the apple of the eye and *capable of feeling outside the limits of the body*.

b. This induces us to point out that the above-mentioned ex periments cannot be classified as part of our domain, psycho-physiology. This field presupposes a conception of the sensitive organ, of the nervous transmission and of brain functions, which is neither necessary nor conceivable. Through experimentation, a Ptolemy is contented with understanding how the sense of sight, which is coextensive to the visual cone of the left eye, combines with the sense of the right visual cone to give the *hegemonicon* power of the soul a unique impression of direction and distance. If we want to call it psychology, then we must think it as the beginning of the old psychology of soul faculties: but not as a premonition of our modern experimental psychology of perception, for which the anatomy and the physiology of the body are an indispensable and unavoidable background.

And yet, this archeo-psychology is experimental.

II. EXPERIMENTATION ON REFLECTION AND REFRACTION

Experimentation on reflection and especially on refraction would lead to the same conclusions: this is an authentic experimental method, but dealing with thought objects and with a distribution of knowledge that have no equivalent in our culture.

1. *Equipment, procedures and results*

Ptolemy has left a detailed description of his equipment. First, he had a metal disc made, divided into four quadrants fully graduated in degrees (fig. 7).

A *dioptron* - an ancient sighting apparatus - (LE) could rotate around a pin E.

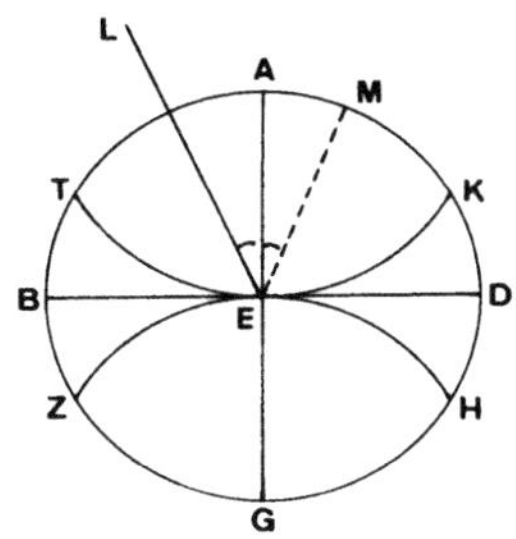

fig. 7

To study reflection, Ptolemy uses three polished metal plates: a flat one, BD, the other two curved into convex mirror ZH and concave mirror TK. By operating successively with the three mirrors pinned on E, he can notice, by moving a sighting mark M and by trying to see it from L, that in the three cases <LEA = <AEM: the angle of incidence of the visual ray is equal to the angle of reflection .

For refraction, Ptolemy uses the same graduated disc. To draw up his table, he moves *dioptron* LE by ten and ten degrees, and in the opposite quadrant (GED) changes the place of the sighting mark M until it becomes visible (fig. 8).

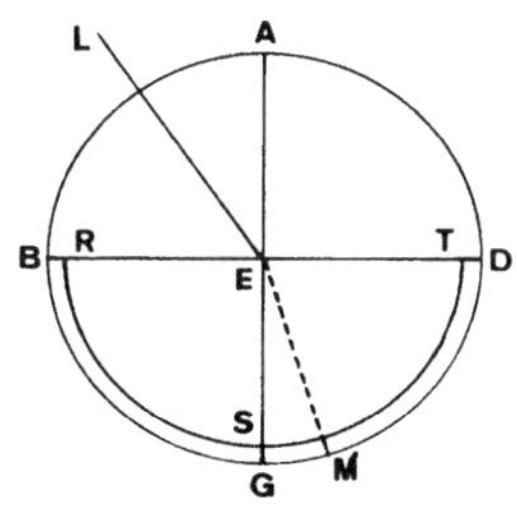

fig. 8

The refraction angle of the visual ray <GEM can immediately be read. For refraction in glass he has moreover built a glass half-cylinder R S T, with a diameter slightly smaller than the quadrant, of which he puts the flat face tangent to B E D.

The result of the tables he draws up for refraction from air to water, from air to glass, from water to glass are satisfactory, since (with the exception of the major incidences, equal or superior to 80 degrees) in all but one case the error is inferior to 1 degree; and the progressiveness of the refraction angle is clearly marked and observed. But if we are dealing indeed with a case of methodical experimentation, this does not mean that this experimentation is made in a scientific field identifiable with one of ours.

2. *The subject of the experiment*

Let us recall that Ptolemy experiments on the *visual ray*. Contrary to what could be thought, it is no way a reversed light ray, although the principle of the reverse return of light provides significance to Ptolemy's experimental findings in both hypotheses.

The case of reflection is a case in point. For a Ptolemy, in a room where there is a mirror, but where there is no one to look, nothing related to vision can happen in or on the mirror. His experiments do not relate to physical phenomena, like propagation of light, which does not depend on the presence of an observer; but to something inconceivable in our culture: a quasi-organ out of our body, proceeding to feel things from a distance according to geometric rules; or an auto-feeling (for direction and length) conic projection of our soul, springing from our pupil, and which evidently does not exist when we are away. Ancient optics is not at all a precocious branch of our mathematical physics.

We once again find the idea that in antiquity, our concept of reflected or refracted *image* does not exist. What can be seen in a mirror or at the bottom of a vessel filled with water is the *very thing*, since the sight actually reaches it. But it can be seen where it is not, with a colour, size or a shape which is often not its own. What is thus seen has therefore the status of an *error of the sense (fallacia visus)*, caught in the trap of a reflecting or refracting surface. At the beginning of Book III, Ptolemy announces very clearly that he will keep on studying the errors of sight (already classified and generally explained in the end of Book II) by going deeper into the errors which occur when the visual rays are broken by reflection or refraction. And *never* during his study of mirrors and dioptres (Books III, IV, and V) does Ptolemy use the word image: but words like "the thing to be seen" *(res videnda)*. Translators or interpreters here add a concept ("image") which is not in Ptolemy's text. The only purpose of these books is to determine *where and how the very thing can be seen* when we look into a mirror or a dioptre. Ptolemy believes that a point reached by a reflected or a refracted visual ray can *always* be seen at the intersection of the ray (produced if necessary) and of the perpendicular dropped from the point to the reflective or refractive surface. In short, this is for him the rule which governs the visual error when the sight is deviated from its natural course.

As a technical consequence, there is no way in such optics to think about improving the vision through what deceives it, reflection and refraction, and making on their base scientific instruments.

CONCLUSIONS

That thought object which is the visual ray is not therefore a reversed light ray. Similarly, ancient optics is neither a forerunner of our psychology of perception (for it ignores the body) nor a forerunner of our mathematical physics (for it does not deal with matter). It is *sui generis*, and belongs to a field of research long since gone. Yet this optics is authentically experimental.

To what field of research is it linked then? To a quite archaic one: the great speculation of antiquity on the relations between sense and sensible, which led to the question of the value of sensitive knowledge. This optics is the part the mathematician played in the investigation and assessment of our senses. It can be called a fossil science, belonging to a classification of knowledge quite forgotten today: a branch of mathematics dealing with sensation and soul.

In his book entitled *Harmony*, Ptolemy explains why mathematicians can deal with such matters. Hearing and sight are the two major senses, for they make it possible for the soul to understand the harmony of sounds and that of heavenly movements. As it is related to ether and light, the soul can be sensitive to intrinsic mathematical proportions of what is purely audible, music, and of what is purely visible, heavenly harmony.[6] The dignity of sense is that of its proper sensible (ἴδιον αἰσθητόν); and the higher it is, the more it can be mathematically studied. Optics is the science of the better sense of the soul, as astronomy is that of the divine part of the world.

We are far from a nature *wholly* written into a mathematical language. This ancient world is organized into a hierarchy, only the most dignified objects of which can be ruled by mathematical relations. As opticians, neither Kepler, nor Scheiner, nor Snell, nor Descartes, nor Galileo himself were better experimenters than Ptolemy, or perhaps than Archimedes four centuries before him. Or (surely) than Ibn-Al-Haytham, who lived in the tenth century. But (like the latter) they knew that they experimented on pencils of light rays, and better than he did, they took into consideration the anatomy of the eye and had at least an inkling of the functions of the nerves and of the brain. The classical scientific revolution is not due to the invention of the experimental method, but to new thought objects, falling within a new distribution of knowledge pattern. Scientific conceptions and events cannot be separated from the whole culture in which they can develop and occur.

Université de Lille III

NOTES

1. Ptolémée. *L'optique de Claude Ptolémée dans la version latine d'après l'arabe de l'émir Eugène de Sicile*. Introduction et notes. Texte établi par A. Lejeune.
2. Lejeune, Albert. *Euclide et Ptolémée, deux stades de l'optique géométrique grecque*. Louvain, 1948.
   Recherches sur la Catoptrique grecque. Mémoires de l'Académie Royale de Belgique, 2ème série, classe des Lettres, 1957.
3. One can find this thesis more extensively developed in our forthcoming book, *Le regard, l'être et l'apparence dans l'optique antique*. Paris, Le Seuil.
4. On the problem of the exact meaning of *aequalis*, see A. Lejeune, *Recherches ...*, Chap. VI, p. 165 et sq.
5. Experiment restored by A. Lejeune (the text was corrupt).
6. See De Pace, Anna. "Elementi aristotelici nell' Ottico di Claudio Tolomeo", *Rivista critica di storia della Filosofia*, Firenze, 1981, 36-2, pp. 123-138.

HULDRYCH M. KOELBING

NEWTON'S AND GOETHE'S COLOUR THEORIES - CONTRADICTORY OR COMPLEMENTARY APPROACHES?

Before we tackle our main theme, let us remember some ideas of Plato and Aristotle. Perhaps not Newton, but certainly Goethe was well aware of a historical continuity ever since antiquity. The Third Part of his Theory of Colours - in the German original "Die Farbenlehre" - is devoted to its history because, as Goethe explains in the Preface of the work, he thinks "that the history of science is science itself", - "daß die Geschichte der Wissenschaft die Wissenschaft selbst sei."[1]

In one of his latest works, *Plato* lets a stranger from southern Italy, *Timaios*, give an account of the creation of the universe and man. As an early and essential step in the formation of our organism, the Gods endowed it with the faculty and organs of *vision*, "the light-bearing eyes", according to the following principle:[2]

> "They caused the pure fire within us, which is akin to that of day, to flow through the eyes in a smooth and dense stream; and they compressed the whole substance, and especially the centre, of the eyes, so that they occluded all other fire that was coarser and allowed only this pure kind of fire to filter through. So whenever the stream of vision is surrounded by day light, it flows out like unto like, and coalescing therewith it forms one kindred substance along the path of the eyes' vision, wheresoever the fire which streams from within collides with an obstructing object without. And this substance, having all become similar in its properties because of its similar nature, distributes the motions of every object it touches, or whereby it is touched, throughout all the body even unto the Soul, and brings about that sensation which we now term 'seeing'."

This is the concept of our seeing by "synaugeia" - a "united radiation" from within and without. There must be a fundamental affinity between the eye and the visible objects, and seeing in a combined action of both. This idea, as elaborated by Plato, goes back to the earlier natural philosophers, especially Empedocles

D. Batens and J. P. van Bendegem (eds.), Theory and Experiment, 189–206.

(5th c. BC); it was taken up by Plotinos, the philosopher in late Antiquity, and strongly appealed to Goethe as a poet as well as a keen observer of man and Nature, in the "siècle des lumières":[3]

> "If the eye were not sun-like,
> it could never see the sun;
> if God's own energy were not within us,
> how could we delight in divine things?" -
>
> "Wär nicht das Auge sonnenhaft,
> die Sonne könnt'es nie erblicken;
> läg nicht in uns des Gottes eigne Kraft,
> wie könnt' uns Göttliches entzücken?"

In a later chapter of "Timaios", Plato gives a hypothetical explanation of the nature of colours: they are sensations, caused by fire, which emanates from the objects, when it is mitigated by the moisture of the eye. Plato goes into some detail about the mixture of colours, but all this, he warns the listener or reader, is hypothetical and of no real importance. In deep earnest, however, he adds:[4]

> "But should any inquirer make an experimental test of these facts, he would evince his ignorance of the difference between man's nature and God's -how that, whereas God is sufficiently wise and powerful to blend the many into one and to dissolve again the one into many, there exists not now, nor ever will exist hereafter, a child of man sufficient for either of these tasks."

A pious resignation, "heilige Scheu", which left its trace, as we shall see, with Goethe. Plato's formula "to blend the many into one and to dissolve again the one into many" sounds to us, of course, like an anticipation of Newton's spectral theory, but the context does not bear out at all such a supposition. The problem is just, I think, colours arise from different components in various amounts.

For *Aristotle*, on the other hand, painstaking investigation of the world in which we live and of the things that are around us is a good and perfectly worth-while occupation, even for a philosopher. The study of Nature yields more information and more accurate knowledge than the favorite speculations about eternal things.[5]

Aristotle worked out *a new theory of sensory perception in general and of vision in particular*,[6] which, however, did not find many followers in Antiquity. To Aristotle, "perception is a form of being acted upon" - thus, it is an essentially passive process. Vision is achieved when the eye is struck by a movement originating from a visible object and transmitted by a transparent medium - air or water. In this way, the eye receives the immaterial forms of the visible objects, "just as the wax receives the impression of the signet-ring without the iron or the gold".[7] This explains, too, the persistence, for a short time, of *after-images* in the eye.[8]

Colour is, for Aristotle, the specific object of vision. In order that colours become actually visible, *light* is needed; light makes the medium between objects and eye actually transparent so that the coloured objects may act through it on the eye and be perceived by it.

As to the differences of colour, Aristotle seems no less at a loss than his former teacher Plato.[9] All colours have their places between the extremes of white and black, between light and darkness. This idea has been taken up again by later students of the subject including Goethe. Aristotle also noted a phenomenon to which Goethe, in his time, was to attribute the highest importance: the effect of a turbid medium on the appearance of colours.

> "The sun appears white when seen directly, but red when seen through fog and smoke".[10]

This common observation, stressed by Aristotle, was to become one of Goethe's "primordial phenomena" ("Urphänomen").

Altogether, it was Goethe's opinion that the Ancients already possessed all the elements of a comprehensive theory of colours (which, of course, would have been much like his own). But unfortunately, they did not manage to purify and unite their empirical observations. Interesting enough, Goethe sees the main reason for this failure in the lack of experimentation. "Experiments (Versuche) are the mediators between Nature and notion, Nature and idea, notion and idea".[11]

NEWTON

So much for the remote background of the more recent discussions and investigations on vision, light and colour. Aristotle had opened the road along which, more than a thousand years later, Alhazen, Averroes, Kepler and Newton proceeded. Alhazen and Kepler may be regarded, respectively, as the most important forerunner and as the true founder of modern visual optics. Speaking of Alhazen, or Ibn al-Haytham (965 to 1035) *Matthias Schramm* said:[12]

> "Something we do not find in Greek literature became here reality: the integration of the anatomical and the physical approaches in the investigation of the structure of the eye."

Alhazen achieved this integration by introducing *refraction of light rays* into his explanation of how an image of visual objects is formed in the eye. But another 600 years went by before John Kepler showed exactly how this is accomplished by the refractive media of the eye. The clue to his discovery was that Kepler, used to experiments in the *camera obscura*, accepted the possibility that an *inverted image* might be formed in the eye, which still would be correctly interpreted by the *sensus communis*, the perceptive faculty situated in the brain. Right down to that time, it had been an undisputed tenet - undisputed even by Leonardo da Vinci - that of course, the eye must receive an upright image of the outer world. As to the nature of colour, Kepler speculated that colour might be a kind of potential light - a feebler light hidden in transparent as well as opaque things.[13]

With Newton, an essentially new insight into this mystery was acquired - an insight that we all are acquainted with since our school days. But it is still interesting to look at the way by which Newton made his discovery! Thanks to his own relation we can follow his observations and reasoning step by step. His report is contained in a famous letter, written on February 6, 1672 and published in the *Philosophical Transactions* on February 19 of the same year already. Remark the astoundingly short delay between the writing and the printing - 13 days!:

> "Sir,
> To perform my late promise to you, I shall without further ceremony acquaint you, that in the beginning of the Year 1666 (at which time I applyed myself to the grinding of Optick glasses of other figures than *Spherical*,) I procured me a Triangular glass-Prisme, to try therewith the celebrated *Phaenomena* of *Colours*. And in order thereto having darkened my chamber, and made a small hole in my window-shuts, to let in a convenient quantity of the Suns light, I placed my Prisme at his entrance, that it might be thereby refracted to the opposite wall. It was at first a very pleasing divertisement, to view the vivid and intense colours produced thereby; but after a while applying myself to

consider them more circumspectly, I became surprised to see them in an *oblong* form; which, according to the received laws of Refraction, I expected should have been *circular*. [...] Comparing the length of this coloured *Spectrum* with its breadth, I found it about five times greater; a disproportion so extravagant, that it excited me to a more then ordinary curiosity of examining, from whence it might proceed."[14]

Thus, what roused Newton out of his peaceful contemplation was not the main feature of the spectrum (the appearance of colours), but a less obvious peculiarity: the distortion of the spot of light by the prism. He had expected to see the colours, but he was surprised to see the image of a circular hole transformed into a bar, five times longer than large. His first suspicion was that this might be an accidental phenomenon, due perhaps to the longer or shorter path of the light through the glass of the prism, or to the size of the hole in the window shutter or the position of the prism either inside or outside - but all this made no difference.

"Then I suspected, whether by any unevenness in the glass, or other contingent irregularity, these colours might be thus dilated. And to try this, I took another Prisme like the former, and so placed it, that the light, passing through them both, might be refracted contrary ways, and so by the latter returned into that course, from which the former had diverted it. For, by this means I thought, the *regular* effects of the first Prisme would be destroyed by the second Prisme, but the *irregular* ones more augmented, by the multiplicity of refractions. The event was, that the light, which by the first Prisme was diffused into an *oblong* form, was by the second reduced into an *orbicular* one with as much regularity, as when it did not at all pass through them. So that, what ever was the cause of that length, 't was not any contingent irregularity."[15]

Not a "contingent irregularity" it was, but a regular phenomenon, an expression of a still unknown natural law. Subsequent experiments and measurements made this more and more clear. At one moment, Newton

"began to suspect, whether the Rays, after their trajection through the Prisme, did not move in curve lines, and according to their more or less curvity tend to divers parts of the wall. And it increased my suspicion, when I remembered

> that I had often seen a Tennis-ball, struck with an oblique Racket, describe such a curve line."[16]

This presupposed that

> "the rays of light should possibly be globular bodies,"

a view widely held at that time, but Newton soon satisfied himself that they continued their path in straight lines also after passing through the prism. And Newton goes on:

> "The gradual removal of these suspicions, at length led me to the *Experimentum Crucis,* which was this: I took two boards, and placed one of them close behind the Prisme at the window, so that the light might pass through a small hole, made in it for the purpose, and fall on the other board, which I placed at about 12 feet distance, having first made a small hole in it also, for some of that Incident light to pass through. Then I placed another Prisme behind this second board, so that the light, trajected through both the boards, might pass through that also, and be again refracted before it arrived at the wall. This done, I took the first Prisme in my hand, and turned it to and fro slowly about its *Axis*, so much as to make the several parts of the Image, cast on the second board, successively pass through the hole in it, that I might observe to what places on the wall the second Prisme would refract them. And I saw by the variation of those places, that the light, tending to that end of the Image, towards which the refraction of the first Prisme was made, did in the second Prisme suffer a Refraction considerably greater then the light tending to the other end. And so the true cause of the length of that Image was detected to be no other, then that *Light* consists of *Rays differently refrangible,* which, without any respect to a difference in their incidence, were, according to their degrees of refrangibility, transmitted towards divers parts of the wall."[17]

"*Light consists of rays differently refrangible*", or, as Newton presently puts it:
"*Light itself is a heterogeneous mixture of differently refrangible rays*[18]", which means in our own terminology: ... of rays of different wave length.

After a digression into the superiority of the reflecting over the refracting telescope, Newton goes on to state his new *Doctrine* thoroughly in 13 propositions. I should like to emphasize four points:

1. "Colours are not *Qualifications of Light*, derived from Refractions, or Reflections of natural Bodies (as 'tis generally believed,) but *Original* and *connate properties*, which in divers Rays are divers. Some Rays are disposed to exhibit a red colour and no other; some a yellow and no other, some a green and no other, and so of the rest."[19]

2. "The Colours of all natural Bodies have no other origin than this, that they are variously qualified to reflect one sort of light in greater plenty then another."[20]

3. "To the same degree of Refrangibility ever belongs the same colour, and to the same colour ever belongs the same degree of Refrangibility. The *least Refrangible* Rays are all disposed to exhibit a *Red* colour, and contrarily those Rays, which are disposed to exhibit a *Red* colour, are all the least refrangible: So the *most refrangible* Rays are all disposed to exhibit a deep *Violet Colour*, and contrarily those which are apt to exhibit such a violet colour, are all the most Refrangible. And so to all the intermediate colours in a continued series belong intermediate degrees of refrangibility. [...] The species of colour, and the degree of Refrangibility proper to any particular sort of rays, is not mutable by Refraction, nor by Reflection from natural bodies, nor by any other cause, that I could yet observe."[20a]

4. After discussing mixtures of colours, Newton, in propositions 7 and 8, comes to his definition of white light, or "whiteness":

> "But the most surprising, and wonderful composition was that of *Whiteness*. There is no one sort of Rays which alone can exhibit this. 'Tis ever compounded, and to its composition are requisite all the aforesaid primary Colours, mixed in a due proportion.
> Hence therefore it comes to pass, that *Whiteness* is the usual colour of *Light*; for, Light is a confused aggregate of Rays indued with all sorts of Colors, as they are promiscuously darted from the various parts of luminous bodies."[21]

So, here we are: pure white light, the light by which the sun illuminates our world - just "a confused aggregate" of coloured rays! This was a paramount scientific discovery, which Newton, starting from a casual observation, had reached by rigorous analytic experimentation and impeccable reasoning. But at the same time, this new concept was a reversal of everybody's immediate sensory experience. This made Newton's theory an unbearable provocation for a mind like Goethe's, for whom sensory perception was the unshakeable basis of all true knowledge of Nature.

GOETHE

Right from the beginning, Goethe's investigation into the nature of colours was determined by his rejection of Newton's experimental and scientific analysis and the spectral theory resulting from it. This rejection was based on Goethe's faith in man's own immediate perception. This comes out clearly in "the Author's Confession" at the end of the Historical Part (=Part III) of "Theory of Colours". He had borrowed a set of prisms, but never used them; at last, their owner, having claimed them back without success, sent a messenger to collect them at once. (This was at the beginning of the year 1791, in Goethe's 42nd year). Goethe accordingly fetched the case which contained the prisms in order to hand it over -

> "[...] when it occurred to me, I might just have a quick look through a prism - something I had not done since my early youth. I remembered that everything appeared many-coloured, but I had forgotten in which way. I happened to be in a completely white-washed room; putting the prism before my eyes and remembering Newton's theory, I expected to see the whole white wall gradually coloured and the light returning from it split into so many coloured lights.
> But how great was my astonishment, when the white wall, viewed through the prism, remained as white as before, while only in places where something dark bordered on it a more or less definite colour appeared. Finally, the window bars showed the most vivid colours, while outdoors, on the light grey sky, not a trace of colour could be seen.
> No long reflection was needed before I realized that, in order to produce colours, *a border* is necessary. At once, and as by instinct, I said aloud to myself that the Newtonian theory was wrong."[22]

In this way, the points were shifted once for all. Throughout the rest of his life, which was to last more than another 40 years, Goethe stuck to this basic conviction of his. Not one of his numerous experiments shook his conviction of the unity of white light.

> "Spaltet nur immer das Licht! Wie öfters strebt ihr zu trennen,
> Was euch allen zum Trutz Eins und ein Einziges bleibt."

This is a poetic version of his creed. But otherwise his polemic against Newton's resembles more the sermon of a fanatic priest against a dangerous heresy than a piece of scientific controversy. A lucid commentator, the late *Andreas Speiser,* dryly states with regard to Goethe's criticism of Newton's "Opticks" that Goethe misunderstood nearly everything.[23] But this is, fortunately, not the whole story.

Before looking closer at the "Farbenlehre", let me stress one point. *Apparently,* Goethe started from the same point as Newton 125 years before: the refraction of daylight by a prism. But he used the prism in a different way and looked, as it were, in a different direction. Newton turned away from the prism, looked at the *objective image* it produced on the wall of a darkened room and saw the spectrum. This, as we have heard, had been Goethe's original intention too, but he did not realize it. What he really did, in his hurry, was to look himself through the prism, in broad daylight towards a white wall or the grey sky. Naturally enough, he could not see a spectrum; but he was struck by the coloured margins along the borders of the clear areas. Without further thinking, he accepted this *subjective perception* as valid evidence against Newton's carefully established discovery. Goethe's whole theory of colours is actually *a theory of colour perception*, and as such it is enormously rich in good observations, and experiments too.

Goethe did not at all reject or condemn *experiment* in itself. He calls it a mediator between object and subject ("der Versuch als Vermittler von Objekt und Subjekt")[24] or a mediator between Nature, notion and idea.[25] He stresses the importance of repeating and varying experiments:

> "The multiplication of any single experiment is the true duty of a scientist."[26]

No less a man than *Helmholtz* admitted that Goethe's own optical experiment were properly done and well described, but as to Goethe's conclusions he qualified them, not as physical explanations, but as metaphorical representations ("bildliche Versinnlichungen des Vorgangs").[27]

Goethe, in the words of his biographer *Emil Staiger*,[28]

> "was annoyed with the extinction of colours amidst refractive angles and figures, with the *translation* of phenomena into mathematical language."

Now, let us come to Goethe's actual observations and teachings! What, in the first instance, *is colour?* For us, thanks to Newton, colour is light; all visible light is essentially coloured light. Goethe did not accept this identification:

> "Colours are acts of light, acts and passions[29]"

he says poetically in the *Preface* of the "Farbenlehre", and more scientifically, he explains a little later in the *Introduction,*

> "that light and darkness, brightness and obscurity, [...] light and non-light are necessary to the production of colour".
> - "Colours throughout are to be considered as half-lights, as half-shadows."[30]

In Goethe's location of colour between the extremes of light and darkness, we still feel the presence of Aristotle. Also much like Aristotle, Goethe states:

> "Colour is an elementary phenomenon in nature adapted to the sense of vision" - "ein elementares Naturphänomen für den Sinn des Auges".[31]

Goethe distinguishes three classes of colours:

> 1. *Physiological colours,* belonging to the eye itself and depending on an action and reaction of the organ;
> 2. *physical colours*, which are perceived in or by means of colourless media;
> 3. *chemical colours*, belonging to the objects.
> "The first are fleeting and not to be arrested; the next are passing, but sill for a while enduring; the last may be made permanent for any length of time."[32]

Each class is discussed in one of the six sections of the Didactic Part (Part I) of the "Farbenlehre". We concentrate on the first two, and I think it is logical to speak first of Goethe's *physical colours* - the colours that are actually produced by refraction.

By the way, it is most rewarding to repeat Goethe's basic experiment (mentioned before) and to look through a three-edged glass prism. With the refracting edge below, any bright surface with darker borders will show a red-orange-yellow lining above and a violet-blue lining below. If this surface is narrow enough - e.g. a strip of white paper - the two colour bands will - as Goethe would have it - meet in the middle and form the full spectrum, yellow and blue producing between them an additional green line.

I shall not try and summarize Goethe's somewhat fantastic explanation of this phenomenon. The essential point is this: Spectral colours are produced by an *interaction of light and darkness*. With the help of a prism, they may be seen to originate at the borders between brightness and obscurity; they may be produced also - and here Goethe becomes perfectly realistic - by an impure optical medium, "ein trübes Mittel"; *Eastlake's* translation of this term is: a semitransparent, or a thickened medium. Paragraphs 150 and 151 of "Theory of Colours" contain Goethe's vivid description of this important observation:[33]

> "The highest degree of light, such as that of the sun, of phosphorus burning in oxygen, is dazzling and colourless: so the light of the fixed stars is for the most part colourless. This light, however, seen through a medium but very slightly thickened, appears to us yellow. If the density of such a medium be increased, or if its volume become greater, we shall see the light gradually assume a yellow-red hue, which at last deepens to a ruby-colour.
> If on the other hand darkness is seen through a semi-transparent medium, which is itself illumined by a light striking on it, a blue colour appears: this becomes lighter and paler as the density of the medium is increased, but on the contrary appears darker and deeper the more transparent the medium becomes: in the least degree of dimness short of absolute transparence, always supposing a perfectly colourless medium, this deep blue approaches the most beautiful violet."

Goethe was sure to have discovered in this pair of observed facts the *primordial phenomena* in the realm of physical colours - "die Urphänomene".[34] Andreas Speiser has summarized them as follows:[35]

> "An illuminated turbidity in front of darkness results in blue to violet; a turbidity in front of brightness, on the other hand, results in yellow to red. On a clear day, Nature presents both these phenomena splendidly to our eyes. The blue sky shows brightness in front of black, sunrise and sunset with their reddish hue show turbidity in front of brightness. Always, Goethe contemplated them with reverence."

As to the value of these observations, *Speiser* - himself a mathematician - stated in 1949:[36]

> "Goethe is right inasmuch as many phenomena known to everybody have not been explained at all by Newton: the blue sky, the red of dawn, the blue or brown colour of the same smoke depending on the background being dark or bright, and so on. It took more than 200 years until they became scientifically comprehensible, while Goethe was able to deduce them without any difficulty from his primordial phenomena. These are still to-day by far the best rules for such colours as appear in normal light."

On the other hand, until the very last weeks of his life, Goethe struggled in vain with the problem of the rainbow, "this embodied Newton" (*Speiser*).

For Goethe, intuition of the primordial phenomena represented the highest revelation of the rules and laws of Nature, the last degree of insight that man possibly could reach. Scientists as well as philosophers should recognize and accept this. Goethe joins with Plato when he says:

> "May the scientist leave the primordial phenomena stand in their eternal silence and glory!"[37]

But this actually means that human understanding goes no father than sensory perception goes and that scientific investigation, including experimentation, must stop here. This, of course, is not acceptable for a modern scientist, though he may admit, like *Werner Heisenberg*, that there is a divine order in Nature.[38]

Goethe's most important and solid contribution to the science of colours is, in my view, the first section of the Didactic Part of the "Farbenlehre", dealing with *physiological colours*, i.e. the subjective colour phenomena. Goethe states:[39]

> "We rightly place these colours first, because they belong altogether, or in a great degree, to the subject, the eye. They constitute the foundation of the whole doctrine and open to our view the chromatic harmony on which so much difference of opinion exists. They have been hitherto looked upon as non-essential and casual, as illusion and infirmity [...]."

But in reality, *they "belong to the healthy eye" and are "necessary conditions of vision."*[40]

Thus, vision in general, and colour vision in particular, depends on the sensory organ as well as on the appropriate exterior stimuli. It is the Platonic and Aristotelian legacy, modernized by Goethe.

To his discussion of physiological colours, Goethe adds an appendix on pathological colour vision, including a description of partial *colour blindness* (daltonism). He is convinced that

> "pathological phenomena point also to organic or physical laws."[41]

By the way, it is on this principle that, at the beginning of our history, Hippocrates has founded the science of medicine, see "The Sacred Disease".

In presenting his "physiological colours", Goethe proceeds systematically. He starts with remarks on the different sensitivity of the retina in darkness and bright light respectively and goes on to the perception of black and white images - "a dark object appears smaller than a bright one of the same size"; after discussing bright-dark contrast, he investigates after-images - and here colour comes in.

With particular zeal, he describes the fact that vividly coloured images leave in our eyes after-images in *complementary colours* (Goethe does not bother to distinguish between ocular and cerebral activity: "the eye" stands, in his text, for visual perception in its totality). The term of complementary colours refers to Goethe's concept of the *colour circle*, comprising the six principle colours from yellow to orange, red, violet, blue, green and yellow again. The colours that stand opposite each other in this circular arrangement form *three pairs of complementary colours*: yellow-violet, red-green, blue-orange. This concept allows Goethe to understand the *coloured after-images* as well as the phenomena of *simultaneous colour contrast* and *coloured shadows*. He likes to exemplify all these phenomena and the natural rules guiding them by observations of his own, the descriptions of which are just delightful. I shall give you two of them. First an instance of *after-image*:[42]

> "I had entered an inn towards evening, and, as a well-favoured girl, with a brilliantly fair complexion, black hair, and a scarlet bodice, came into the room, I looked attentively at her as she stood before me at some distance in half shadow. As she presently afterwards turned away, I saw on the white wall, which was now before me, a black face surrounded with a bright light, while the dress of the perfectly distinct figure appeared of a beautiful sea-green."

I cannot help to admire Goethe's visual sensitivity as well as his readiness to record so exactly the enchanting appearance, to analyse and to interpret it at once! This is in perfect accordance with a statement of his from the *Preface*:

> "Every act of seeing leads to consideration, consideration to reflection, reflection to combination, and thus it may be said that in every attentive look on nature we already theorize" - "[...] daß wir schon bei jedem aufmerksamen Blick in die Welt theoretisieren."[43]

According to the principle of complementary colours, he explains also the *colour of some shadows*:[44]

> "In travelling over the Harz in winter, I happened to descend from the Brocken towards evening; the wide slopes extending above and below me, the heath, every insulated tree and projecting rock, and all masses of both, were covered with snow or hoar-frost. The sun was sinking towards the Oder ponds. During the day, owing to the yellowish hue of the snow, shadows tending to violet had already been observable; these might now be pronounced to be decidedly blue, as the illumined parts exhibited a yellow deepening to orange.
> But as the sun at last was about to set, and its rays, greatly mitigated by the thicker vapours, began to diffuse a most beautiful red colour over the whole scene around me, the shadow colour changed to a green, in lightness to be compared to a sea-green, in beauty to the green of the emerald. The appearance became more and more vivid: one might have imagined oneself in a fairy world, for every object had clothed itself in the two vivid and so beautifully harmonising colours, till at last, as the sun went

> down, the magnificent spectacle was lost in a grey twilight, and by degrees in a clear moon-and-starlight night."

In the fact that we see after-images, simultaneous contrasts or shadows in colours complementary to those of the primary stimuli, Goethe saw a natural law, a *law of vision*:[45]

> "In these instances, the eye essentially demands totality and completes in itself the colour circle."

In the description I have just quoted, the cooperation of the objects seen and the subject looking at them is expressed, I think, in a particularly beautiful way.

Newton and Goethe - contradiction or completion? In comparing their theories of colour, we actually find both aspects.

Goethe contradicted fanatically Newton's discovery regarding the objective nature of spectral colours and, of course, he was wrong. Astonishing enough, he even realized, at least at times, that he was prejudiced. At the beginning of the second, the Polemical Part of "Farbenlehre", he tells us blandly what he thinks of the value of experimental proof:[46]

> "All opinions about things belong to the individual, and we know only too well that conviction depends not on insight, but on volition; that nobody comprehends anything unless it suits him and, therefore, he is ready to admit it. In knowledge as well as in action, prejudice decides everything, and prejudice is, as its name says, a judgement before investigation. It is an affirmation or a negation of something that appeals to our nature or conflicts with it; it is a joyful urge of our vital nature towards the true as well as the wrong - towards everything with which we feel in harmony."

Thus, Goethe's conviction of the unity of white light, and of colour as being derived from light and darkness jointly, was a prejudice, and he knew it. But he wrongly assumed that Newton's spectral theory, in spite of its experimental proof, was a prejudice too. His intuition made him believe that his own prejudice was true while Newton's was not.

But, as I earlier said, this is not the whole story. Goethe was the first modern author to insist so strongly on the *active participation* of our eyes, and implicitly also our brain, in vision, the first to investigate it so thoroughly and to describe, systematically and carefully, the subjective conditions of colour vision. (He presented a psychology of colour perception, too, in the 6th section of the Didactic Part of "Farbenlehre".) In a masterly way he stated the case of the sensitive subject in visual perception. In this respect, Goethe's approach to the phenomenon of colour was, and still is, an indispensable and essential complement to the experimental scientist's investigation as represented by Newton.

Institute for the History of Medicine
University of Zürich, Switzerland

NOTES

1. AGA *16*, 13.
2. Timaios, 45B–D, Loeb 101/3.
3. In slightly different wording in: Farbenlehre, Didactic Part, Introduction, AGA *16*, 20.
4. Timaios, 67C–68D; Loeb, 173–177.
5. De partibus animalium *1*, 5, 644b–645a; Loeb, 97/99.
6. De anima, and De sensu et sensilibus. See *Koelbing*, Das Problem der visuellen Wahrnehmung in der Antike, im Mittelalter und in der Renaissance, *Fortschr. Ophthalmol., 80* (1983), 11–14.
7. De anima *2*, 11/12, 424a; Loeb, 135–137.
8. De anima *3*, 2, 425b; Loeb, 147.
9. We are indebted to *Marie-Hélène Marganne* for a learned exposition of Atristotle's chromatic system: Le système chromatique dans le Corpus Aristotélicien, *Les études classiques, 46* (1978), 185–203.
10. De sensu *3*, 440a (in Parua Naturalia); Loeb, 235.
11. Farbenlehre, Part *3* (Historical), AGA *16*, 329–31.
12. *Schramm, Matthias*, Zur Entwicklung der physiologischen Optik in der arabischen Literatur, *Sudhoffs Archiv 43* (1959), 288–316. My translation.
13. Quoted from *Goethe*, AGA 16, p. 421f.
14. A Letter of Mr. *Isaac Newton*, Mathematick Professor in the University of Cambridge; containing his New Theory about *Light* and *Colors* [...], *Philosophical Transactions*, Numb. 80, February 19, 1671/72, 3075–3087. Quotation from p. 3075f. At that time, in England the new year was still counted from March 1st, therefore, 1671 is given as well as 1672. The latter corresponds to our Gregorian calendar.
15. p. 3076.
16. p. 3078.

17. p. 3078f.
18. p. 3079.
19. Proposition 1, p. 3081.
20. Proposition 13, p. 3084.
20a. Propositions 2 and 3, p. 3081.
21. p. 3083.
22. *Goethe*, AGA *16*, 708f., My translation.
23. *Speiser, Andreas,* AGA *16*, 944.
24. AGA *16*, 844.
25. AGA *16*, 331.
26. AGA *16*, 852.
27. *Helmholtz, Hermann von;* Handb. der physiolog. Optik, 3. Aufl., vol. *2*, Hamburg/Leipzig 1911, p. 96.
28. *Staiger, Emil*, Goethe *2*, Zürich 1956, reprinted 1970, 421.
29. AGA *16,* 9.
30. TC lvif.; AGA *16*, 22f.
31. TC liv; AGA *16*, 22.
32. TC lv; AGA *16*, 23.
33. TC 61f; AGA *16*, 63.
34. Para. 175; TC 72; AGA *16*, 69.
35. *Speiser, A.*, in AGA *16*, 942.
36. in AGA *16*, 970.
37. Para. 177; AGA *16*, 69.
38. *Heisenberg, Werner*, Das Naturbild Goethes und die technisch-naturwissenschaftliche Welt, in: Die Grossen der Weltgeschichte, vol. 7, Zürich 1976, p. 66-73.
39. Para. 1; TC 1; AGA *16*, 27.
40. Para. 3; TC 2; AGA *16*, 27.
41. Para. 102; AGA *16*, 53. See *Wolfgang Jaeger*, Goethes Untersuchungen an Farbenblinden, in *Heidelberger Jahrbücher 23* (1979), 27-38.
42. Para 52; TC 22; AGA *16*, 39.
43. Preface; TC xl; AGA *16*, 11.
44. Para. 75; TC 34f; AGA *16*, 47.
45. Para. 60, AGA *16*, 47.
46. *Goethe*: Zur Farbenlehre. Polemischer Teil. Bearb. von *R. Matthaei*. Weimar 1958. Para. 30: "Beweis durch Experimentieren", S. 12.

Abbreviations mean:

AGA: *Goethe, Joh. Wolfgang*, Artemis Gedenk-Ausgabe der Werke, Briefe und Gespräche, Zürich 1949.

Loeb: Loeb Classical Library, London/Cambridge, Mass. Volumes quoted:
Aristotle , On the Soul/Parva Naturalia/On Breath, 1936, revised and reprinted 1957.
Aristotle, Parts of Animals [...], 1937, revised and reprinted 1961.
Plato VII, Timaios [...], 1929, reprinted 1961.
TC: *Goethe, Joh. Wolfgang*, Theory of Colours, translated by *Charles Lock Eastlake*, London 1840, reprinted by M.I.T. Press, Cambridge, Mass./London, 1970.

ARISTIDES BALTAS

ON THE STRUCTURE OF PHYSICS AS A SCIENCE*

INTRODUCTION

No one doubts that the historicist turn in Philosophy of Science has considerably shaken – to say the least – the way logical empiricism pictured science. The problem areas and the particular questions that this turn brought with it and the older ideas it shed its own light on (theory change, scientific progress, theory ladenness, the Duhem-Quine thesis and its consequences, incommensurability etc.) created a wholly new image within which the traditional distinctions (e.g. context of justification/context of discovery, internal/external history) became blurred, if not explicitly and thoroughly erased. However, as it usually happens in Philosophy, the problems and questions which made up the historicist movement and constituted its force were many times posited and reposited, they were amply discussed from many different angles but, after all, they were not "solved" (or dissolved) – at least in a way that did built up a substantial consensus. The net result is that disillusionment has replaced the excitement and the high hopes that the historicist movement created in its heyday. Thus, the pendulum is swinging back towards more general and more traditionally philosophical issues: Realism, rationality, new variants of empiricism attract more and more the attention of practitioners in the field.

To my mind, however, the efforts to understand the different sciences in all the complex and diverse aspects of their structure and of their history – above all that of Physics – should not be abandoned. At least not yet. I have three main, separate but interrelated, reasons to support this exhortation which may add up to the fundamental tenets of a long term research program.

The first reason consists in a general philosophical appraisal of the historicist movement as a whole. I state it here axiomatically and in two words, without the long arguments it certainly necessitates: The historicist critique of the logical empiricist picture of science was, in my consideration, mainly local; it left the fundamental philosophical assumptions of the empiricist picture practically unchallenged. Thus it could not help remaining

D. Batens and J. P. van Bendegem (eds.), Theory and Experiment, 207–225.

confined within the space delimited by these assumptions, within what one may roughly call modern anglo-saxon empiricism. And since these assumptions determine the questions that can be posed and the answers considered acceptable, one may infer that these assumptions are responsible for the stumbling blocks and the dead ends of the historicist movement. Therefore, it is legitimate to say that a new philosophical look is required a philosophical look willing and ready to tackle the premises and presuppositions usually taken for granted or considered (mostly implicitly) as beyond dispute.

The second reason amounts to an article of faith. At least as far as Physics is concerned, I believe that it is possible to critically assess the different insights that the historicist movement has offered in order to arrive eventually at a comprehensive and empirically oriented theory of the structure and of the history of that science. When I qualify the wished-for theory as "empirically oriented", I basically have in mind a theory the claims of which can, in one way or another, be put to the test of what physicists do and of what they come up with. As this theory is supposed to come out as the comprehensive theory of the structure and of the history of the *science* of Physics, the very process of its own construction should provide us with some indications as to what a theory should in general look like for a test of its scientific status to be meaningful. Such indications or criteria should also be applicable to the theory generating them without giving rise to any vicious self referential circle.

The third reason brings closer to earth what I maintained as my first reason and somewhat tempers the exorbitant ambition barely concealed in the second. It constitutes the claim that the starting point as well as the basic ingredients of the theory, the aims of which I just, very roughly, described, *actually exist.* They are to be found at the points where two wildly different approaches to the phenomenon of sciences paradoxically converge. The first such approach is that of the French marxist philosopher Louis Althusser. The second is the approach introduced by Joseph Sneed, expounded by Wolfgang Stegmüller and further elaborated by Wolfgang Balzer and Carlos-Ulises Moulines, among others. I am well aware, of course, that this claim is most controversial if it does not sound down and not crazy. However I will not defend it here.[1] For reasons of economy, I will proceed by immediately proposing my synthesis of these two approaches as regards the particular case of Physics. In this way the picture of Physics I will try to depict can be assessed in its own right, independently of how fairly I treat the two approaches that I consider as grounding this picture.

According to me this synthesis amounts to the following thesis: *Physics is a structured whole made up from three constitutive elements that mutually define each other - its conceptual system, its object and the experimental procedures specific to it.* Each of those three elements is not composed and put into place independently of the other two; each is what it is and performs the epistemic function that it performs only through the interrelations that tie it from birth to the others and which the unique process that establishes all three sets up from the start among them. Moreover, it is these interdefining relations which engender - in the modes of interplay they institute among these three elements - the different facets of that science's dynamics. The present paper consists in an effort to spell out this thesis.

The paper is divided into two parts and a conclusion, each part being subdivided into two sections. The first part describes how the conceptual system of Physics is established through a process that works on natural phenomena as these are given before Physics' constitution as a science. The first section here considers this system as an already established abstract structure and discusses what I call the systemic component of the meaning of the corresponding concepts. The second section discusses, on the one hand, how this system is effectively established and, on the other, how, in the process, a second meaning component - the empirical component - is assigned to these concepts. The second part concerns the process through which the concepts and relations making up the conceptual system of Physics are "attributed" to phenomena. This process is in fact, inseparable from the one discussed in the first part and only expository reasons oblige me to treat each separately. Here the first section discusses the "attribution" of the concepts and how Physic's reference domain is delimited while the second section refers to the attribution of the relations and discusses what the experimental procedures specific to Physics amount to and how they are put into place as well as how Physics' object is finally constituted. As it will be noticed, Althusser and Sneed are barely mentioned in what follows. This omission is, again, due to economical reasons and does not imply that I "forget" what I owe to their respective approaches. To state the point as unambiguously as possible, the present paper does not pretend to be anything more than a "synthetic comment" on what these two approaches have already established.

A. FROM NATURAL PHENOMENA TO PHYSICS' CONCEPTUAL SYSTEM

A.1 *The Systemic Component of Meaning*

The conceptual system of Physics (in the present paper I will concentrate my discussion on and draw examples mainly from the "foundational component"[2] of this system, namely the conceptual system of Classical Mechanics) is made up from a set of concepts and of the relations that tie these concepts together. These concepts are the foundational as well as the derivative concepts of Physics (position, time, instantaneous velocity, acceleration, material point, mass, force etc.), while these relations are mathematical relations, independently formulable in the terms of some particular mathematical theory (or theories) which "underpins" the conceptual system in question (e.g. Vector Calculus for the case of Classical Mechanics)[3]. The whole system constitutes a structure in the strict, technical meaning of the term: each element of the system (each concept) is what it is only in virtue of the relations that the system itself as a whole institutes between it and its other elements, only in virtue of its particular *position* within the system. At this level of analysis therefore, Physics' conceptual system is abstract and contentless. The meaning of each physical concept (or, rather, the *systemic component* of this meaning, for we will discuss in a moment the second, equally essential component of this meaning) is determined solely by the intra-systemic relations that define its position within the conceptual system. For example, the concept of "force" acquires the systemic component of its meaning through its intrasystemic relations to the concepts of "material point", "instantaneous velocity", "mass" etc.

On the above view, the systemic component of meaning of a physical concept is not determined fully by the relations of this concept to the set of concepts which, in a deductive exposition of the system, serve to define it and which, in this sense, the concept in question presupposes. The intrasystemic relations of this concept to the set of concepts which it itself contributes in defining and which, in this sense, it entails are also implemental in the determination of its meaning. The systemic component of, say, the concept of "momentum" is determined by both the set "material point" "mass", "instantaneous velocity" etc. and the set "impulse", "conservation of momentum" etc. The position of the concept within the conceptual system is thus to be conceived as a knot in a multidimensional net and therefore such position possess an "invariant" or "absolute" (in respect to the different possible deductive chains that lead to and start from it) character. It is such determination of the systemic component of meaning by this invariant or absolute position which allows these concepts to conserve their meaning through different formulations of the same conceptual system – e.g. the Newtonian and Lagrangian formulations of Classical Mechanics; conversely, it is this conservation which allows such formulations to be considered as different formulations of one and the same conceptual system.

As it is evident, the conceptual system of Physics is not established once and for all in the totality of its possible ramifications. The system initially contains a limited set of concepts in a limited net of their interdefining relations. But, on the other hand, as this system is a structure, it entails consequences. Theorems can be proved, new concepts can be defined, distinctions can be introduced, a whole complex network of secondary concepts and relations can be produced[4] (e.g. the Galilean transformations, the concepts of momentum, work, kinetic and potential energy). This developability of Physics' conceptual system enriches the systemic component of meaning of all of its concepts. A concept's position within the system becomes defined by more specifications and, consequently, it acquires new aspects of systemic meaning.[5] As we will see in what follows, this flourishing of the systemic component of meaning through the deployment of the conceptual system's consequences plays a crucial role in our "measuring" the distance that separates the way Physics accounts for the phenomena making up its object from the way common sense apprehends such phenomena.

A.2 *The Empirical Component of Meaning*

The conceptual system of Physics is not, of course, a free creation, an arbitrary game of gratuitous concepts that float unrestrained above reality. If it is to be a physical conceptual system then the concepts and relations that make it up must have something to do with the natural phenomena that Physics cognitively accounts for. To see how this is actually the case we must first turn to the way Physics' conceptual system becomes effectively established.

The phenomena that are to make up Physics' object in no case are given in the raw. Most of them are phenomena that people encounter in carrying out their diverse social practices and with which, in one way or another, they have to cope if these practices are to be reproduced. Different arrangements of notions, images, representations etc. come thus about by means of which these phenomena become "socially appropriated", that is apprehended and treated (through appropriate attitudes, gestures, behaviors etc.)

in such a way that the social practices in which these phenomena are encountered can effectively follow their course. Such arrangements of notions - images - representations in their mutually constitutive association with such attitudes - gestures - behaviors make up what Althusser calls *practical ideologies*.[6] Particular practical ideologies are thus shaped by the particular social processes that make up social practice as a whole while, in return, those ideologies assure the cohesion of these processes by providing them with the ideas and stances that their reproduction requires. The reproduction of these processes within social practice as a whole then validates their corresponding practical ideologies, "confirms" the ideas (the notions - images - representations) that make up these ideologies and endows these ideas with the rigidity and unquestionableness of the self-evident. Practical ideologies are responsible for all what we innocently and unknowingly accept as given and thus they form the basis of our common sense.

But phenomena are not apprehended through practical ideologies alone. Althusser distinguishes practical from *theoretical ideologies* and considers the latter as theoretical systematizations which try to account for particular domains of reality as these domains are being actually treated by the existing social practices and their corresponding practical ideologies. Theoretical ideologies sanction - or perhaps contest - such practices and such practical ideologies by offering "reasons" or by "explaining" them but they always share and they never question or reveal what the avowed premises, procedures and results of such practices silently admit as given or what they take tacitly for granted. Theoretical ideologies thus warrant such given in its effects while this given, in return, contributes its self-evidence as their evidential support and as the grounds on which they seat their authority.[7] In this way theoretical ideologies not only inform but also consolidate common sense. It is the interplay between practical and theoretical ideologies which provides our understanding of a domain of phenomena before the science that is to account for these phenomena enters the scene. The process which leads to the establishment of Physics' conceptual system should, therefore, be conceived as a process that works on and transforms in some radical manner the notions-image-representations of the practical ideologies (as they have been variously elaborated by the corresponding theoretical ideologies) which treat the phenomena that are to make up Physics' object.

In particular, this process[8] works on such notions - images - representations, makes them cast off an essential part of the connotations, import and implications with which the ideologies in which they are involved link them, connects them in new ways modifying them correspondingly and finally detaches them from their ideological environment by transforming them into elements of the abstract conceptual system we have been discussing. For example, from the notions with which practical and theoretical ideologies came to terms with phenomena related to the change of position of different objects and to the different perceivable characteristics of such changes, this process produced the new concept of "instantaneous velocity". Concurrently, the same process divests real objects from all their multiple and diverse characteristics with which practical and theoretical ideologies distinguished and identified them and arrives at the formulation of the new concept of "material point" as the "possessor" of instantaneous velocity. At the same time all such characteristics condense and disappear into the new concept of "mass", the sole qualitative and quantitative feature of material points. The conceptual system thus produced is abstract in the sense that the process which establishes it transforms away the ideological determinations that make up the concrete content of the notions on which this process worked.

In fact, this detachment from ideology becomes completed through the development of the conceptual system's consequences. With this deployment the concepts of the system acquire aspects of meaning which the practical and theoretical ideologies from which the system originated find impossible to expect, conceive or control. Classroom experience shows that common sense continues even today to receive as a paradox the proposition that, if we disregard air resistance, a ton of lead and a chicken feather will arrive simultaneously on the ground if they are left simultaneously to fall from the same height. The "measure" of the distance separating common sense notions from physical concepts is thus the "measure" of the autonomy of the latter in respect to the former.

But however "big" or "small" such an autonomy might prove to be, all the consecutive products of the transformations leading to Physics' conceptual system as well as this system itself *never cease to refer to the natural phenomena they were produced to account for.* As this system, as well as all its predecessors, was devised only in order to account for some such phenomena, a continuity of reference with respect to them is preserved throughout all such transformations with the result that the conceptual system which is finally established remains "anchored" to these phenomena in the sense that it *congenitally talks about them.* In other words, the conceptual system of Physics is an *always already physically interpreted* system and thus the concepts that make it up are endowed from birth with a second component in their meaning: the meaning of a physical concept is not only systemic; *it is simultaneously systemic and empirical.*

But on the basis of what we have said so far, a natural question may at this point come to mind: Even if one may grant that *some* of the concepts included in Physics' conceptual system might well possess the empirical component of their meaning which the physical interpretation of that system assigns to them from birth, what about the concepts which are produced by the deployment of that system's consequences *independently of any predetermined reference restrictions?* To this question I will answer in the next section by showing that the relations which Physics' conceptual system entertains with the phenomena it effectively accounts for possesses, among its other features, the power to render all the derivative concepts of that system referable to aspects of these phenomena and, correspondingly, to endow these concepts too with an empirical component in their meaning.

But although each physical concept derives its meaning from two different sources - namely from its position within the conceptual system on the one hand and from its "anchorage" to the natural phenomena it refers to on the other - it constitutes one and only one entity the two-front identity of which is conjointly determined by both these meaning components at once. Each concept has, as it were, two faces resembling in that respect to a coin. For example "force" is both the cause of "acceleration" and what we exercise when we try to push some body; and "acceleration" means simultaneously both the "derivative of instantaneous velocity over time" and what we sense when our car speeds up.

As it should be evident from what precedes, the empirical component of meaning can never completely break off its ties from ideology and become fully detached from it.[9] This ideological character of the empirical component of meaning creates a permanent tension between the two meaning components. On the one hand the systemic component tends to fix the overall meaning of the concept by trying to tie it fast to the clear cut position that defines it within the discrete array of knots and mathematical relations among knots that make up the abstract conceptual system; on the other hand, the empirical component interprets the concept and, by the continuous spectrum of uncontrollable connotations that this interpretation brings with it, tends to fill in the interstices between such knots and, correspondingly, to blur their distinctness. When, for example, Newton's second law is interpreted by saying that force causes acceleration, the notion of cause surreptitiously introduces a swarm of ideological allusions, nuances and overtones which the abstract conceptual system tries to check and to restrict but never manages to master fully. Even the notion of motion, by "naturally" implying that position and velocity can "of course" be simultaneously determined, carries with it the host of ideological commitments from which Quantum Mechanics, will, precisely, have to extricate itself. The two meaning components can thus never completely collapse on one another thereby rendering the conceptual system "perfectly" scientific. The coin representing each concept always keeps a certain, scientifically indeterminable, "thickness".

The persistence of this tension between the two meaning components constitutes a salutary feature for it assures that such ideological assumptions and commitments will eventually be revealed, that this conceptual system will correspondingly become reinterpreted and that the science of Physics will continue to develop. In fact, we here encounter the second essential aspect of Physics' dynamics, the first being the developability of its conceptual system independently of any reference restrictions.

B.—FROM PHYSICS' CONCEPTUAL SYSTEM TO PHYSICAL PHENOMENA

B.1 *The Delimitation of Physics's Reference Domain*

The process which establishes Physics' conceptual system is not restricted to the one way direction we have been concentrating on up to now. The reverse, as it were, direction of this process, that which *reattributes* the concepts and relations produced to natural phenomena plays an equally essential role and demands in its turn our attention. Needless to say that these two opposite ways of operating are not in any way separate. The process in question consists in a single movement that goes in both these opposite directions at once.

In its "backward" operation, the process under discussion reattributes, as we just said, the concepts and relations of the *ex hypothesi* already established conceptual system to natural phenomena transforming them thereby into physical phenomena. The attribution of that system's *concepts* first (only expository reasons make us distinguish the two cases) engages each single phenomenon by itself and leads to the delimitation of Physics' reference domain; the attribution of the *relations* making up that system, then, always takes in its stride a set of phenomena and leads to the constitution of the domain of Physics' effective applications. In the present section we will examine the first aspect of this process of attribution; in the next section the second.

Before embarking on this, however, I think I have to point out that the term "attribution" is intentionally chosen here in order to demarcate the present view from views considering that the appropriate term in this context should rather be that of "abstraction". For my part, however, I cannot admit the idea that the process I am trying to describe consists in an operation that abstracts (or should we say "extracts"?) Some kind of innermost essence or essential part that somehow already exists "inside" these phenomena and lies hidden there, as it were. To my view the concept, say, of a "material point with a determinate mass" does not constitute the common essence of apples, planets and projectiles. It is rather a concept of the conceptual system of Physics which is *produced* together with the other concepts of this system and which is, precisely, ***attributed*** to such real objects so that their movement may be accounted for by this system as a whole.

The "backward" operation works on the phenomenon itself: It erases almost all the multiple facets that make up this phenomenon as empirically given, it strips it of all the properties and features by which practical and theoretical ideologies distinguish and individuate it and account for it and it picks up its single particular aspect which is idiosyncratic to the conceptual system which is being simultaneously established. Newton's apple is not an apple anymore. It is not the offspring of an apple tree, it is not a fruit, it does not possess skin, kernel or seeds, it does not have colour, taste or smell. It is only something that falls. It is to this particular singled out aspect of the natural phenomenon, it is to this "something that falls", that the concepts of the system are attributed, conceptually transforming in the process that aspect too. What really falls is not, in fact, just something; it is a "material point". And the real fall of this material point is neither some indeterminate change of position nor the "natural" movement that makes this point or body return to its "natural" place of rest at the center of the universe. It is the accelerated motion brought about by the action of a force; it is the "free" (or, rather, coerced) fall due to gravitational attraction.

The way the concepts of Physics' conceptual system acquire the empirical component of their meaning becomes thus clearer. The abstract concepts of "material point", "force" etc. by becoming attributed to a natural phenomenon not only conceptually transform this phenomenon but *become themselves* the empirical concepts involved in its treatment. The concepts produced in the "forward" operation are, in the very process of their production, *immediately* reattributed to the phenomena therein involved. *As the natural phenomenon becomes conceptually transformed the concepts become empirically meaningful.*

This transformation is completed only after the *whole* phenomenon has been totally reduced to just its thus selected and transformed aspect. Such a phenomenon constitutes a *physical phenomenon.* The net effect of the operation we are discussing is to transform *a natural phenomenon into a physical one*.[10]

A physical phenomenon is, in a sense, a *new* phenomenon. The conceptual transformation of the phenomenon *by just the foundational concepts* assures that the *derivative* concepts of the system can also become attributed to this phenomenon thereby licensing the transmissibility of empirical meaning from the foundational to the derivative concepts. The concept, say, of "momentum" is a derivative concept defined through the concepts of "mass" and "instantaneous velocity". And since after its conceptual transformation and reduction, the *real* movement of a *real* billiard ball is *nothing but* a material point or body with determinate mass and determinate instantaneous velocity, the concept of momentum can be attributed to *that real billiard ball* and thence acquire its anchorage to reality which endows it with the empirical component of its meaning.

It follows now that the operation we are discussing defines some kind of an equivalence class among natural phenomena. After the falling apple has been transformed into a material point that falls because of gravitational attraction, a pear, a stone or a raindrop, in fact *everything* that falls, can become a "material point" that falls for the same reasons. However, such a universality in attributability is not specific to Physics' conceptual system. In fact practical ideologies do treat phenomena as the above as empirically similar and theoretical ideologies ground themselves on this empirical similarity and account collectively for them with the notions they elaborate for that purpose. The difference between a theoretical ideology and a scientific conceptual system as that of Physics lies precisely in the fact that, for the latter, such a universality in attributability is *not* restricted to empirically similar phenomena. As the practical and theoretical ideologies determining what such an empirical similarity amounts to are precisely the ideologies from which this conceptual system has been detached, these ideologies can have no say as to which phenomena the concepts of the system can in fact be attributed. The attributability of this system can thus be extended to *empirically independent phenomena,* that is to phenomena such that the experience of each does not involve the experience of the others. In this way, as empirically independent phenomena come under the jurisdiction of Physics' conceptual system, the ideological frontiers that distributed such phenomena into empirically separate domains are destroyed, the ideological bonds that held these phenomena together within such domains are broken and new distinctions, new qualifications, new ontological realms are established. The equivalence class of phenomena to which Physics' conceptual system can be attributed becomes thus *a new reality domain*, the reference domain of Physics' conceptual system.

The ability of Physics' conceptual system to refer to empirically independent phenomena adds a further dimension to Physics' dynamics. From what we have said so far it follows that no immanent restriction exists to prevent the concepts of that system to become attributed to more phenomena and to conceptually transform them in their turn so that they become included too into that system's reference domain. Physics' conceptual system has thus the inborn ability to extend its scope without limits thus keeping the frontiers of that science's reference domain continuously open, always ready to receive new members.[11]

But the above does not imply as yet that Physics' conceptual system does indeed cognitively account for the phenomena thus encompassed into Physics' reference domain. The last section formulates an answer to that problem.

B.2 *The Experimental Procedures Specific to Physics and the Constitution of Physics' Object*

What needs to be clarified is how Physics can proceed in order to check whether the *abstract* relations that make up its conceptual system do or do not hold for any particular *material* phenomenon of that system's reference domain. For this to be possible, what is needed in the first place is a standard with respect to which such a check can be performed. And this standard should possess both the above heterogeneous features at once: It should both be something material – that is something with which a *material* phenomenon can be related – and it should also be something *abstract* – that is something for which *abstract relations* make sense. Does such a thing exist? The obvious candidate is a physical phenomenon, as we have defined it; it is both a material phenomenon happening in that material world out there and a phenomenon conceptually transformed by the concepts of Physics' conceptual system and conceived exclusively in their terms. But although any of them can in principle play the role of such a standard, a further requirement, is that *the relations* that make up Physics' conceptual system *are already satisfied* by it.

So *let us for the moment suppose* that, for some reasons, this is indeed the case for a given physical phenomenon. Then this phenomenon, under conditions and possible restrictions expressly specified by Physics' conceptual system, can be "frozen" into the form of a *device or apparatus*.[12] This device or apparatus is constructed so as to be employed in a series of practical, material moves - themselves also explicitly prescribed by the same conceptual system - which make up a particular procedure "measuring" if the relations of the abstract conceptual system hold for *another* phenomenon, in general empirically independent from the "frozen" one. The "frozen" phenomenon thus makes up the "standard of comparison", the "gauge" which the phenomena included into Physics' reference domain have to match up with; and it is the match between this "standard of comparison" and another physical phenomenon which constitutes the trial through which each such phenomenon has successfully to pass if it is to become an effective application of Physics: Only the positive outcome of such a measurement procedure turns the physical phenomenon from a potential[13] into an effective application of that science.

To take a simple example, let us suppose that the relations which Physics' conceptual system attributes to the physical phenomenon of pendulum movement are satisfied. Namely, that the "small" oscillations (as this qualification is expressly specified by the same conceptual system)- of all material pendulums with equal length are isochronous. Then such a material pendulum can be "frozen" into the form of *a clock*; and this clock can then be employed in the particular material procedure which the same conceptual system prescribes as composing, in respect to this particular "standard of comparison", the procedure measuring time intervals in "free" fall. The results of such a measurement procedure fulfill what Physics' conceptual system expects from them and thence the physical phenomenon of "free" fall can effectively become, in conjunction with the physical phenomenon of pendulum movement, a member of the domain of Physics' effectively applications. It is in such material correlations among physical phenomena that the experimental *procedures specific to Physics* consists. And it is the *effective* installation of such material correlations among different physical phenomena which builds up the domain of Physics' effective applications out of the referenced domain of that conceptual system. Thus, what passes judgement as to if a phenomenon can rightfully become an effective application of Physics is *another phenomenon* to which the corresponding conceptual system already refers, a phenomenon which has already been conceptually transformed and thus rendered "homogenous" with the phenomenon under test, a phenomenon for which the relations that make up the conceptual system already hold. The search for an absolute foundation is replaced by a *test of mutual compatibility*. And as nothing guarantees beforehand the positive outcome of such a compatibility query and that such phenomena can in actual fact become thus materially correlated, the effective refutation of the claims constituted by the relations making up Physics conceptual system is always possible. Therefore the judgement that such experimental procedures pass remains unbiased in respect to the conceptual system that supports them. This judgement is indeed a fair, impartial judgement.

The above show clearly why the transformation of a physical phenomenon into an effective application of Physics cannot engage a single phenomenon in isolation. *The transformation of a physical phenomenon into an effective application of Physics always presupposes the completion of this transformation for another.*[14] Therefore no phenomenon can become by itself Physics' *first* effective application. The process which – in its "forward" operation – established Physics' conceptual system and simultaneously – in its "backward" operation – reattributes the concepts and relations of the system to natural phenomena arrives at a point where *a set* of such phenomena *prove together* that they can become mutually materially correlated in the sense that *each* one of them guarantees that the relations which make up Physics' conceptual system hold for *all* the others. These phenomena form together the set of *founding members* of the domain of Physics' effective applications. And as we know, the domain of effective applications of Classical Mechanics was historically founded by "free" fall, by pendulum movement, by planetary motion and by tidal phenomena.

These founding members possess the privilege to lay down the rules by which the phenomena included into Physics' reference domain should abide if they are to become effective applications in their turn. Thus it is the effective exercise of this privilege which structures and organizes the whole of Physics' reference domain thereby constituting *the object of Physics*: Physics' object is nothing by the reference domain of that science *as structured by its effective applications,* it is these effective applications *together with* the physical phenomena which are in the process of becoming effective applications in their turn.

CONCLUSION

As it should by now be clear, the "constructivist" picture on the structure of Physics that emerges out of the proceeding pages is based on assumptions different from those grounding the corresponding logical empiricist picture in any of the latter's numerous versions. Physical phenomena do not just lie there, ready to be explained by some theory which is devised to that end by some physicist's cleverness or genius. These are phenomena that are *constructed* out of the natural phenomena which our experience – as it is formed and informed by practical and theoretical ideologies

- encounters. And the particular claims that such a theory formulates are neither the bold conjectures that our physicist's cleverness musters out of the blue nor are they tested by some device which remains neutral in respect to them and which the cleverness, again, of the same or another physicist manufactures for the purpose. Physics' conceptual system, its object and the experimental procedures specific to it are neither given *a priori* nor manufacturable at will independency of one another. They are *all produced together* within a unique process which composes each in relation to the others, which assigns to each its particular function in the cognitive appropriation of the part of reality that these three elements carve out together and which, moreover, fixes the modes of their interplay, the modes that make Physics develop further. Thus, instead of the two level picture that logical empiricism generically posits - namely that of a set of propositions making up a theory and that of a theory-free experience, each level being practically autonomous in respect to the other, relating to it only through independently introduced "bridge principles" (correspondence rules, reduction sentences etc.) - the present view speaks of three constitutive elements that interdefine each other. And instead of a dialogue between two radically different parties - some concepts and some propositions containing them on the one hand and some phenomena experienced independently of such (or even of any) concepts on the other - the present view considers Physics as structured by a whole complex network of interrelations which closely interweaves and simultaneously confronts concepts with concepts, concepts with phenomena and phenomena with phenomena.

At first glance at least, the logical empiricist picture of Physics may appear far simpler than the above and intuitively more appealing. But the second look with which the recent evolution in Philosophy of Science has provided us has convincingly shown - at least to me - that the traditional picture is in effect much too simple. Moreover, if one entertains the understandable ambition to arrive somehow to make out all - as far as it is humanly possible - the extremely complex and diverse facets which the structure and evolution of the science of Physics has shown to possess - and this as coherently, as rigorously, as economically as possible - then one should be ready to face this complexity.

Physics Department
National Technical University, Athens

NOTES

* Most of the work that led to the present paper was done at the Center of Philosophy of Science of the University of Pittsburgh. It is a great pleasure to thank from here Nick Rescher for his warm and multiform hospitality in this fine institution he directs as well as his assistant, Linda Butera, who made everything easy for me there. Preliminary work related to the paper and earlier versions of it were presented at some of the "lunch time talks" of the Center as well as at Virginia Tech in Blacksburg, VA and at Union College, Schenectady, N.Y. My sincerest thanks go to all those who participated in the seminars and helped me with their comments and criticism. Dionysios Anapolitanos, John Beverley, Kosta Gavroglu, Paul Humphreys, Vasso Kindi, Tom Nickles and Jean-Paul Van Bendegem privileged me with a close reading of previous versions of the paper: their comments led, I am happy to say, to some drastic changes. In addition, Jean-Paul Van Bendegem contributed decisively in rendering manageable the length of the final version. Peter McLaughlin expressed his interest in my approach as a whole and forwarded a series of very helpful comments. Finally, Peter Machamer, Enrique Eduardo Marí, Alexander Nehamas, Pandelis Nicolakopoulos and Marcello Pera not only read the paper and commented copiously on it but also helped me and encouraged me in ways my thanks cannot ever hope to do justice to.

1. I discuss this claim in some detail in A. Baltas "Louis Althusser and Joseph D. Sneed: A Strange Encounter in Philosophy of Science?" *Criticism and the Growth of Knowledge: Twenty Years After. An International Conference in Memory of Imre Lakatos.* Thessaloniki, Greece, August 1986. To be published.
2. The conceptual system of Classical Mechanics does not constitute the "foundational component" of the conceptual system of Physics from the obvious historical point of view only. It is such also in the sense that it makes up the *conceptual basis* with which conceptual systems that come later in the field have to be articulated (possibly modifying it in the process), if they are to become integral parts of the science of Physics, that is if they are to become, precisely, *physical* conceptual systems in their turn. E.g. it is the material points of Classical Mechanics which by acquiring charge, become sources of the electromagnetic field. Such articulations may not always be logically impeccable. But they form necessary conditions for the continuation of the *actual practice of Physics* as the practice of a *particular science.* Among other

things, this practice has a task to complete and to logically refine such articulations. It is in this sense that E.C.G. Sudarshan and N. Mucunda and their *Classical Mechanics: A Modern Perspective* (John Wiley and Sons, 1974, Introduction, p. VII) consider Classical Mechanics "... not as a part of physics, but as physics itself."

3. A whole lot can be said, of course, on the different aspects of the relation between Physics and Mathematics: one and the same physical conceptual system can be formulated in terms of very different mathematical theories; one and the same mathematical theory can provide the means for the formulation of very different physical conceptual systems etc. Moreover it seems that its privileged relation with Mathematics singles out Physics from among the natural sciences and confers to it its specificity. For these matters see Jean-Marc Lévy-Leblond, "Physique et Mathématiques" in François Guénard and Gilbert Lelièvre, (eds.), *Penser les mathématiques*, Seuil, 1982. For some additional remarks and comments see A. Baltas and K. Gavroglu, "A Modification of Popper's Tetradic Schema and the Special Relativity Theory", *Zeitschrift für allgemeine Wissenschaftstheorie*, XI/2 (1980) pp. 213-237.
4. For some characteristic features of these secondary networks and their relation to Popper's concept of "World 3" see A. Baltas and K. Gavroglu, op. cit.
5. It goes without saying that here I am speaking only of a "mature" physical conceptual system, that is a system which a well defined mathematical theory already "underpins". In this way a "mature physical conceptual system" amounts to what Sneed calls "a theory of *mathematical* physics". However the actual practice of Physics comes up with a "mature" physical system only after a long struggle with what physicists call "models". For a distinction between the two see A. Baltas and K. Gavroglu op. cit. and the references given there. In particular, see R. Feynman, *The Character of Physical Law,* the MIT Press, 1975 and especially Chapter 2.
6. For a definition of the concept of "practical ideology" see L. Althusser, *Philosophie et philosophie spontanée des savants,* Maspéro, 1973, p. 26.
7. *Ibid*., p. 64 and *passim*. Under this definition the Aristotelean theory of motion is an example of a theoretical ideology. It is a theoretical construction which, by construction, preserves what everyday practice and the practical ideologies which structure it present as given, it draws its evidence from such a given and tries to explain it without every questioning it. See also G. Canguilhem, "Que est-ce qu'une idéo-

logie scientifique?" in his *Ideologie et rationalité dans les sciences de la vie*, Vrin, 1977.

8. This process constitutes in fact what Althusser calls - following in that respect G. Bachelard - *the process of an epistemological break.* In this terminology the present paper is nothing but the explication of the *mechanism* that operates such a process in the particular case of Physics; and in this way the title of the paper could very well be "The Mechanism of the Epistemological Break in the Case of Physics".
9. I develop these matters in A. Baltas, "Ideological 'Assumptions' in Physics: Social Determinations of Internal Structures", *PSA 1986*, volume 2. To be published.
10. The aspects of the phenomenon which the conceptual system of Physics ignores and reduces away are not, of course, completely lost forever. Through the operation of similar mechanisms, they make up the object of other sciences or scientific disciplines. For example, although *from the perspective of Classical Mechanics* the real empirical Earth is *just* a material point (or a material sphere) that moves under the influence of the Sun's gravitational attraction and its crust, its interior, its oceans, its earthquakes etc. *do not exist*, these latter aspects of the empirical phenomenon "Earth" are not altogether forsaken; they make up the object of disciplines such as Geology, Oceanography, Seismology etc.
11. It is this innate tendency of a science's reference domain to continuously broaden its horizons which, on the one hand, makes the force of the developing science and, on the other, - under suitable "external" conditions - may authorize ideology to proclaim impudently the "final solution" - mechanical (Laplace), biological (E.D. Wilson), "marxist" (Lyssenko) etc. - of "all" problems. However, from what we have said, it follows that such a tendency is not completely unmitigated. If there exist no inherent limits as to *which* phenomena can fall under the scope of Physics' conceptual system, this system can *only* treat the *single* particular aspect of each phenomenon that is idiosyncratic to this system - and this only by *totally reducing* the whole phenomenon to just this aspect. The *universality* in scope of this system as regards phenomena to which it can be attributed is harnessed, as it were, by its *particularity in the effective treatment* of each phenomenon.
12. This idea forms one of the fundamental tenets of Bachelard's philosophy of science. For him "un instrument dans la science moderne est véritablement un théorème réifié". See G. Bachelard, *Les intuitions atomistiques,* Boivin, 1933, p. 140.

For a general appraisal of Bachelard's philosophy of science from the point of view that concerns us here see Dominique Lecourt, *Marxism and Epistemology,* NLB, 1973. See also Mary Tiles, *Bachelard: Science and Objectivity*, Cambridge University Press, 1984.

13. Following Sneed, what we termed "physical phenomenon" could have been called a "potential application" (of Physics).
14. This idea constitutes one of the fundamental insights of the Sneed approach and follows directly from Sneed's theoreticity criterion demarcating theoretical from non theoretical terms within a "theory of mathematical physics".

SILVANA BORUTTI

MODELS AND INTERPRETATION IN HUMAN SCIENCES: ANTHROPOLOGY AND THE THEORETICAL NOTION OF FIELD

> In anthropological investigation, objectivity lies neither in the logical consistency of a theory, nor in the giveness of data, but in the foundation ('Begründung') of human *intersubjectivity*.
>
> Fabian, 1971, p. 25

The present paper consists of three parts. In the introduction (I), I describe briefly the epistemological model which is the premise of my analysis. My argument is that knowledge in human sciences is essentially *interpretive knowledge*. In the central part of my paper (II), I deal with the problem of defining what interpretation is in a particular human science, that is anthropology. Once assumed, as I do, that an anthropological fact is only given in relation to a point of view, or to a pre-comprehension of the fact, we must analyze the specific features of such a relation. The framework of the cognitive relation, in the ethnographic experience, is the *field*. Given that, the field is not to be taken as a background or a generic container of this relation. Rather, it determines its specific features. As we shall see, these features are basically dialogical. Finally, in the third part (III), I shall put forward some points in order to discuss the role of *language* and *dialogic interaction* in the cognitive strategies of anthropology.

I

Now, let us begin with human sciences as interpretive models. When we talk about the cognitive status of human sciences, we define it as a kind of interpretive knowledge. In the first part of my paper, I shall analyze briefly what interpretation turns out to be in human sciences as far as the problem of knowledge is concerned. There are two points which have to be stressed: (i) firstly, human

D. Batens and J. P. van Bendegem (eds.), Theory and Experiment, 227–241.

sciences consist of models by means of which objects are built up: consequently, objects are not pre-given data, but the result of a symbolic construction; (ii) secondly, the cognitive subject in human sciences is not a paradigmatic instance, an ideal subject of knowledge, but a historical subject embedded in a relation of interpretation.

Let us now analyze these two points separately. First of all, the kind of knowledge in human sciences. Here we cannot rely on objects with properties, or given referents, but we must produce models and concepts for the attribution of properties and meanings. In other words, the cognitive approach in human sciences implies the shaping of objects and the construction of points of view aimed at the construction of reality. This constructive kind of knowledge - which (incidentally) is typical of, but, I consider, not confined to the human sciences - can be defined in several ways: for example, we find 'seeing as' in Wittgenstein (1953), 'seeing that' in Hanson (1958), and in Mary Hesse (1966) we find the expression 'metaphorical redescription'. I think that one of the most interesting ways of indicating this constructive character of knowledge is suggested by Paul Ricoeur (1975). He defines metaphors and models as forms of 'poietic redescription' of reality, that is, a kind of redescription which produces reality while producing an unknown (not yet known) view of reality. Ricoeur derives the word 'poietic' from the Aristotelian concept of *poiesis*. In *Poetics* (1450a and 1451b), Aristotle defines tragedy as a *mimesis* of reality *via mythos* (myth). While history deals with accidental individuals, tragic *poiesis* has the philosophical and theoretical property of detecting ontologies, that is of revealing the essence of human reality by means of the inventive quality of myth. According to Ricoeur, metaphors and interpretive models have the same poietic property as the Aristotelian myth: they do not reproduce, but they invent forms to shape reality. In my opinion, the word *poiesis* perfectly suits the constructive nature of interpretive models in human sciences. What I mean is that such models are 'symbolic' and 'absolute' at the same time. I use 'symbolic' as in Cassirer (*Philosophie der Symbolischen Formen*), in the sense of symbolic ways of worldmaking. And I use 'absolute' as in Blumenberg (*Paradigmen zu einer Metaphorologie*) in the sense of an untranslatable, or nonsubstitutable activity of metaphorical thought.

This radically constructive nature of interpretive models, that I suggest calling *poiesis*, or invention, defines well the circular link which exists between theory and referent in human and social sciences. Let us take an example from economics. It makes a difference to consider facts and data in terms of the economic model of marginal utility (and of the law of supply and demand); or to consider them in terms of the Marxian theory of value (and of the relations between capitalists and workers). We can say that here observation and explanation are not separable: in other words, the hypotheses by means of which we observe the facts (in the example, the society as a market or the society as a class struggle) will constitute the theory itself by means of which we explain the facts. Observation, instead of being a testing of models, is itself a result of these models. Wittgenstein suggests something like that when he says: don't look for the objects, look for the grammars and the games. In *The Blue Book* he says that, instead of looking for that mysterious object, that secret entity, that is the meaning, it is better to ask "what is an explanation of the meaning of a word" (Wittgenstein, 1969, p. 1). This is a way of saying that we have to analyze not the given objects, but the interpretive models. This is the first point to be emphasized in an interpretive approach: that is, what the interpretive approach stresses in human and social sciences is a kind of *constructive* knowledge, which redescribes and invents objects.

The second point I want to underline is that to talk of interpretation in human sciences is also to indicate a particular cognitive relationship between subjects and objects. The Heideggerian and Gadamerian hermeneutics says that in the interpretive situation there is a circular link between subjects and objects. In other words, the understanding is a historical understanding, which takes place through the mediation of language and tradition. It is, according to Gadamer, a pre-understanding or a pre-comprehension: a prior grasping which is not the work of a subjective consciousness, but of language and tradition. For example, nowadays we do not merely interpret the meaning of Renaissance texts; rather, the comprehension of their meaning is a possibility of self-comprehension for men today. There is, so to speak, an 'ontological complicity' between subjects and objects. Kant ascribed this complicity to the *a priori* structures of subjectivity (like, for example, transcendental schematism); at the present time, after the so-called 'linguistic turn', we are more inclined to find it in language and in models. As hermeneutics shows, language and models are the media of our understanding.

Obviously, these media are not a neutral instrument, a method or a meta-language available to the subject. On the contrary, they are laden with the meanings derived from a historical and cultural tradition, from that background of texts called by semioticians 'intertextuality'. To use an analogy of Lévi-Strauss (1962), these media are similar to the pieces that a *bricoleur* collects and keeps in his tool box: in other words, the instruments of our understanding are not the result of an ideal, artificial language, but they are signs determined by their historical origin.

These are, generally speaking, the constituents of an interpretive situation. I think they offer a correct description of the cognitive relationship in human sciences, and particularly in anthropology. I shall turn now to a more specific problem: given that anthropology is a kind of interpretive knowledge, how can we identify the interpretive elements which characterize the relationship between anthropologists and anthropological objects? In other words, what are the elements which define the anthropological objects as symbolic constructions?

II

I shall now discuss how the interpretive approach has been applied in anthropology. Clifford Geertz, the main figure of the so-called 'interpretive turn', says that interpretation of cultures is "understanding of understanding": that is, "an attempt somehow to understand how we understand understandings not our own" (Geertz, 1983, p. 5). Geertz sticks to a semiotical concept of culture, and maintains that ethnographers must look at cultural facts (actions, myths, institutions, rituals) really as *texts* , or, better, as manuscripts:

> Doing ethnography is like trying to read (in the sense of "construct a reading of") a manuscript - foreign, faded, full of ellipses, incoherencies, suspicious emendations, and tendentious commentaries, but written not in conventional graphs of sound but in transient examples of shaped behavior. (Geertz, 1973, p. 10).

According to Geertz, the theory and methodology of interpretation in anthropology can be perfectly described by such an analogy: meaning, in looking at facts as if they were texts. This looking at facts is a kind of understanding from a distance. This is the sense of the analogy: to be an anthropologist is not, in Geertz's words, "seeking either to become natives ... or to mimic them", as romantics or spies would do (Geertz, 1973, p. 13), but rather "to construe [I stress this word] ... their symbol systems" (Geertz, 1983, p. 70). The anthropologist must use the meanings of his own culture to reconstruct the way in which people belonging to a different culture make sense to themselves. To the anthropologist, his meanings work like a 'seeing as' which allows the understanding of the meanings of a foreign culture. For example,

Geertz analyses the State organization in nineteenth century Bali by using the metaphor of a theatre, and he describes the life in Bali as if it were a ritualized 'social drama', where persons have only a public identity (Geertz, 1980).

In such a model of analysis we can find the essential element of the interpretive approach: that is, to connect the meanings of different cultures in a sort of dialogue:

> "Translation", here, is . . . [a] displaying the logic of [others'] ways to putting [things] in the locutions of ours; a conception ... rather closer to what a critic does to illumine a poem than what an astronomer does to account for a star. (Geertz, 1983, p. 10).

This is the reason why the interpretivists pay attention to those models according to which various groups represent their own symbolic behaviours (the so-called 'folk-models'). Finally, in the interpretive approach the emphasis is on the constructive aspect of the anthropological transcription: like a literary or artistic critic, the anthropologist construes, writes in, or inscribes (to use an expression from Ricoeur, 1970), invents the meanings of a social discourse as if it were a literary or visual text. Anthropological writings are thus 'fictions': this does not mean that they are false or unfactual, but, as Geertz says, "something made, fashioned" (Geertz, 1973, p. 15).

This approach is certainly a suitable criticism of that anthropological pattern in which the ethnographer, like an entomologist, "observes, records, analyzes" (Geertz, 1973, p. 20). But in my opinion what we cannot either find or look for in this model of analysis are the *actors* of the social discourse: we can find the utterances of the social meanings, but we cannot identify the enunciative roles of the social actors (that is, the roles of the social actors in making speech acts: 'actes de parole', or better, 'actes de discours', see Ducrot, 1972). We can find the meanings, but not the subjects of these utterances: that is, the natives (those who live by these meanings), the informants (the natives who assist the members of another culture in interpreting these meanings and their language), and finally the anthropologists (those who arrange the conditions for communication and interpretation). There is, I think, a constitutional weakness in the metaphor of the text used by Geertz (see Lowry, 1981; Shankman, 1984). In fact, this metaphor describes the social discourse as if it were a static and limited set of meaning, rather than indicating the process by means of which these meanings are produced. Moreover, this model leaves out of account the interaction between the anthropologist and his informant in the field. This interaction is the stage of inquiry which precedes the anthropological writing, that is, in Geertz's words "our construction of the other people's constructions" (Geertz, 1973, p. 9). What Geertz does not realize is that not only the anthropological text, but also the interaction between anthropologists and informants is responsible for the construction of meanings. At the end, Geertz puts forward once again what Kevin Dwyer calls a 'contemplative model' of the scientific attitude: Dwyer means by that a one-way relationship between a passive observer or recorder on one side, and objects in isolation on the other (Dwyer, 1982, p. 263). In the interpretive reconstruction of the anthropologist, the natives are not speakers, but rather utterances and texts. In other words, the model deals with systems of written symbols.

Thus even in the interpretive approach we find an old methodological presupposition: that anthropological knowledge has to begin with preconstituted forms of rationality (naming, ways of recording concrete data, like questionnaires and synoptic tables), while neglecting the relationship between anthropologists and informants. This is for example Malinowski's methodological prescription (in the *Introduction* to *Argonauts of the Western Pacific*): the fieldworker must get in touch with the natives, but he must also be provided with a "mental chart", that "ought to materialize into a diagram, a plan, an exhaustive synoptic table of cases" (Malinowski, 1922, p. 14). His methodological advice is explicit: a written or graphic arrangement must first of all control and predetermine the communicative relationship between anthropologist and informant in the field, and then transcribe it in the anthropological text. It seems as if the oral exchanges had nothing to do with the theoretical constructions. (Incidentally, in spite of Malinowski's 'participant observation', his diary published after his death reveals the actual behaviour of the anthropologist in the field (Malinowski, 1967): idiosyncrasies and fights with the natives, aversions, self-engrossment, and so on. This was a source of embarrassment among anthropologists at the time. But actually that diary poses an epistemological problem (see Pinxten, 1981): it just reminds us that the field is not an ideal scenario, but a real communicative setting).

Here we come to the main point. An interpretive approach cannot both reject the model of natural sciences, where observation is built up as an experiment in a given environment, and neglect to take into account the communicative relationship between *interpreting* subject and *interpreted* subject in the field. This is precisely what is at stake: the conditions for interpretation of the Other (of the native) are given in the field, before any actual writing. These conditions consist of those communicative strategies that link interpreter and informants in a dialogical interaction. Let us now analyze the nature of this interaction.

As we have already seen, the informant is the leading actor of this exchange. This figure has an epistemological importance. In fact, we cannot think of the informant as confining himself to the task of giving bare information. Why? Because information here consists of a communicative flux, which gradually grows, between anthropologist and informant. What they exchange are not statements. They exchange what Ducrot calls "le dire" (Ducrot, 1973): that is, speech acts, or discursive acts connected in a dialogical process. So to speak, and in Gadamerian terms: the experience of understanding is a dialogical process, since to have a world is to have a language, and "being which can be understood is language" (Gadamer, 1965, p. 450). Actually, the scene of anthropological knowledge starts with the *logos*, with a hard face-to-face speech construction. All the anthropologists say that the field starts with a reversed-roles ritual, the observation of the observer; but they say that incidentally, in the footnotes or in the preface of their books. Nevertheless, what is taken to be marginal is a fundamental scene in order to understand how knowledge begins. There is a sequence of speech acts (including of course the obstacles and breaks in communication), in which the observer and the observed subject ascribe enunciative roles to each other in order to communicate. Few anthropologists give an account of this interaction.

Among these few, is Jeanne Favret-Saada (1977, 1981). Her books are in fact an account of ethnographical work as a dialogic interaction. She investigates the diffusion of the so-called belief in witchcraft and magic in the Bocage (Normandy). Therefore she needs no interpreter of language, but only informants who answer her questions. But she immediately finds out that the informants, once asked about their own involvement with magic, try to avoid any direct reply by using speech devices, like dissimulation. In such a way, they think that they conform to the supposed symbolic system of the ethnographer: "Do you think we are savages?", they say, implying there is no magic at all. But, apart from the obvious fact that trying to break communication is still a way of saying something in a specific communicative context, the gaps and the difficulties in the communicative exchange stand for an implicit but crucial move: *the observation of the observer*. By being reticent, they implicitly ask the ethnographer to declare her own identity and her own role in the communicative context. No informant, either from the Bocage or from Melanesia or from Morocco, is willing to believe that the anthropologist asks for information to no purpose. The first rule that the fieldworker learns from his informant is that there is no neutral collection of information: this can only be a scientific ideal, it is certainly not a possible communicative practice.

Any fieldworker (not only in anthropology, but also in oral history, in sociolinguistics and so on) who deals with informants has stories to tell about initial misunderstandings. The mutual embarrassment turns out to be an exchange of roles as far as observation is concerned. The informants start by putting questions on the identity of the anthropologist, on his origin, on his tribe. For example: "Are you from the government, form the missionaries, form the doctors? Can you anthropologists intermarry with the government and the missionaries?" Roy Wagner is asked by the Daribi of New Guinea (Wagner, 1974, p. 19). Anyway, a symbolic and ritual work through language makes the interference acceptable. Remo Guidieri (1980) recounts that one of the Fataleka aged men (in Melanesia) is sent to meet him in order to decide whether to welcome him or not. The success of the meeting does not depend on the gifts and the other objects shown him, but on a little misunderstanding between the two of them. The anthropologist asks for information about the flute music played in that area, and the Fataleka man understands that *he* is asked to play some flute music. He makes sure that they are alone, and then he improvises some pieces using a bottle. Only afterwards does the anthropologist realize that the native had made for him an imitation of a female song and of a mortuary song. At that time, the anthropologist had no clue to grasp the actual value of the gesture, which was a rule-breaking 'mise-en-scène' of the native's culture (Guidieri, 1980, pp. 25-30).

The observer is kept under observation: he is not allowed to stay outside, looking at and collecting facts while imposing an order on the communicative exchanges with the natives. Returning to Favret-Saada, she writes that she had to face silences and mistrust as long as she persisted in putting questions on magic. This situation changed when she realized that the only way towards understanding was to take up the challenge: following the direction of dialogue, in the end she accepted the linguistic roles made up by the dialogue itself.

The account of Favret-Saada is tense and vivid: she shows us the dramatic structure of her experience. She wants us to know above all that her understanding begins when she does not act any longer as a subject of knowledge outside the field, and she lets the Other (the native) challenge her identity form inside the field. In the words of Lévi-Strauss:

en comprenant et en se faisant comprendre, on livre quelque chose de soi, et ... on prend prise sur l'autre. (Lévi-Strauss, 1949, pp. 615-616).

Favret-Saada says that the crucial moment of her understanding is when she can see that the only empirical fact she has to face in the field is either silence or speech, in both cases *language* (Favret-Saada, 1977, p. 20). Witchcraft itself is a speech, a speech which is neither knowledge nor information, but a power contest: this is what she learns form her informants. In fact, after being silent and reticent, her interlocutors change the communicative rule into a face-to-face confrontation of strengths and powers. They ask the anthropologist: "Are you strong enough to talk about magic?"; and they address her alternately as if she were a supposedly powerful subject (a witch, or a witch-doctor), or as if she were a supposedly powerless subject (a victim, a bewitched person). In such a way, they explain to her that witchcraft is a man-to-man fight, in which only speeches are powerful. The anthropologist understands that there are no statements describing magic, but only possible enunciative roles: what can be said, and what cannot be said, and by whom, and in which circumstances. Who speaks of witchcraft? Only those who are caught up in it, either witches, or witch-doctors, or bewitched people.

This is what happens to Favret-Saada, and the communicative sequence in which she becomes a subject of magical speeches is as follows: (i) some bewitched people tell the anthropologist their stories for the first time once they identify her as a witch-doctor who could help them; (ii) a few months later, a countryman composes in a series a number of accidents which have occurred to the anthropologist, announcing that she is bewitched and taking her to a witch-doctor; (iii) the anthropologist for two years submits the events of her personal life to the interpretation of a witch-doctor; (iv) the anthropologist herself is identified as a witch-doctor by some countrymen, but she fails to take up this role; (v) in the meantime, the relationship between the anthropologist and the witch-doctor becomes more and more complex, since the anthropologist is supposed to be both a patient and a witness of the witch-doctor's powers.

In such a way, Favret-Saada treats the field as a dialogue, as a 'séance d'énonciation'. At the very last, she says that the anthropological understanding and interpretation both begin when the questioning takes into account not the content of the natives' utterances, but their ways of speaking, that is the speech acts performed in the field. Why this silence? What does this negation mean? (I use 'negation' in the Freudian sense of *Verneinung*,

Freud, 1925). Why just this utterance in this context? Who is speaking, and to whom? Wittgenstein would say that the anthropologist, in order to understand, has been trained in a form of life, and that the training, here as in any other society, consists of the taking part in communicative games and their rules. (Incidentally, it is interesting that Wittgenstein used to invent examples of a speculative ethnography: by creating games, fictitious countries, tribes, rules in order to understand our rules and our games, Wittgenstein merely simulated the 'culture shock' (see Wagner, 1975), or the *dépaysement*, which is the condition of anthropological understanding).

III

I shall now summarize the main points I have been analyzing so far. From the account of Favret-Saada's experience we can essentially infer two consequences: (i) knowledge in anthropology starts as an observation in the field; (ii) this kind of observation in the field is structured as a dialogic relationship, that is as a communicative interaction through language. This dialogical pattern of observation in anthropology criticizes another pattern which, following Dwyer, I have called contemplative. The *contemplative pattern* treats anthropological knowledge as a relationship between a subject possessing a method and an independent object. In my opinion, the interpretive anthropology is based on the same contemplative pattern. To the extent that this approach maintains that anthropologists construe texts which are constructions of other people's constructions, the anthropologists' constructions are taken to be practices applied to a passive material. On the contrary, the *pattern of the dialogical fieldwork* that I acknowledge (following mainly Favret-Saada and Dwyer) claims that anthropological knowledge is indeed a kind of interpretation; but that such an interpretation is the result of an experience shared both by the anthropologist and his informant in the field. The main feature of this pattern is the description of knowledge as a contextual and historical event. According to this point of view, the anthropological writing cannot be silent on the dialogical interaction in the field, as if it were a premise that can be left out of the final account. In reality, the fundamental points of view in understanding are built up in the field.

And now, after my brief summary, a few considerations about: (i) the fieldwork as a dialogic event, and (ii) the dialogue as a communicative environment where interpretations are construed.

As to the first point, the fieldwork as a dialogic event is the initial stage of anthropological knowledge. The anthropologist in the field cannot confine himself to observing pieces of behaviour and to recording natives' verbal statements as if these were simply single descriptions of social reality: he must also enter into a dialogical relationship. By producing an account of this relationship, the anthropologist stresses the contextual, strategical and historical quality of his own understanding. As we have seen in our examples, the confrontation in the fields has certain dialogical characteristics on which anthropological understanding depends. In the field interaction the utterances are *sequential*, that is they appear at a definite point of the communicative chain; the utterances are *indexical*, that is they are tied to the space-time context; the interaction is *marked* both by a mutual search for contexts which will make the communication possible, and by conflicts and breaks: in other words, the interaction is not cooperative by definition (as some analyses of the logic of conversation claim), but it can be developed as a mutual manipulation, or as a real fight, by using presuppositions, digressions, dissimulations, meta-communication (discourse about the relationship: Bateson, 1972); finally, the construction of meanings is not a progressive and linear chain, but the result of a particular kind of *rearrangement*; this rearrangement depends on recurrences of words and on aggregations of meanings that only appear at the end of the communicative chain. Following Parret (1984), we can speak of pragmatic orientation of understanding in the field. To assert that these dialogical features of the fieldwork are the basic elements of anthropological understanding is not to claim that the anthropological writing has to be only an exact recording of the dialogues in the field. This is not the point. The point is that the recording has to show the contextual nature of any dialogue, that is, as Dwyer writes:

> the structured inequality of the partners during their encounter: the anthropologist singles out "events" and poses questions; the informant answers, embellishes, digresses, evades. (Dwyer, 1982, p. XVII)

There is an essential challenge in the field between the Self and the Other. That is why the understanding in the field is the main step towards the construction of interpretive models.

Let us analyze the second point mentioned above; it is in the fieldwork, as a dialogic event, that the anthropologist devises his interpretations. In the communicative interaction with the informants, it is the pre-structured understanding and the hypotheses of the anthropologist that induce the informants' answers. And these answers are not a simple mention of models already present in the informants' minds; rather, they are the result of the interaction and clash between the so-called folk-models, with which the natives describe their own social relationships (see Holy and Stuchlik, 1981), and the forms or symbolic schemes offered by the anthropologist's questions. In fact, those speech acts, naming questions and answers, which prove to be essential in the interaction between anthropologist and informant, point out that every move in the interaction results from, and acts on, the other moves (this is shown for example in Michel Meyer's analysis of the questioning process: Meyer, 1986). In any anthropological dialogue, there are at the same time rearrangements both of subjective roles and of the anthropologist's presupposed schemes and hypotheses. For instance, Favret-Saada produced her interpretation of magic in the field in the following way. She started by asking some questions about witchcraft as if it were a form of belief or the kind of knowledge of naïve persons. But she had soon to give up this pattern of questioning, and in order to understand she had to try to get herself involved in a sort of risky social game. At last she understood that magic cannot be treated as a *belief* versus *knowledge*, but it must be understood as a form of world-construction - a form which gives identity to subjects through language and puts events into a meaningful series. Another example to show that not all anthropologist's questions are acceptable questions. Guidieri asks for information about a stone-working technique, and the informant answers by telling him a myth. In fact, to a Fataleka man to explain is not to give technical explanations, but to make a phenomenon 'human', as they say, by putting it into a network of connections by means of ritual speeches. In such a way, the anthropologist realizes that in Fataleka culture only myths are complete explanations, while technical reasons are a sort of maimed explanations. Here the question-answer sequence is fundamental in view of interpretation (see Guidieri, 1980, pp. 401-402).

From this point of view, the dialogue appears as a metaphor of the interpretive feature of anthropological knowledge. According to Dwyer, the transcription of the dialogical event

> must be looked at not as a model but as a suggestive metaphor for an anthropology that embraces its own wager character. (Dwyer, 1982, p. 286)

As a matter of fact, the anthropological text, before reaching the structure of models of social reality, is written by means of the rhetorical and pragmatical devices that are typical of dialogical interaction. So I conclude that anthropology is not interpretive only because it describes and construes its objects as if they were texts to be decoded. Rather, I believe that the interpretive nature of anthropology also depends on the fact that the cognitive relationship is here embedded in the mediation of language and of models of scientific and cultural tradition. Hermeneutics says that comprehension is not a psychological concept, but rather it is a linguistic event determined by our prejudgements (*Vorurteilen*). I think that this perfectly suits interpretation in anthropology. The anthropologist has not to become the Other, but he has to undertake enunciative roles in interaction with the Other; in other words, he has to put himself into a discursive situation. Interpretation, or invention, of the Other's culture already starts from this initial stage, before the anthropologist transcribes the dialogue into forms of indirect discourse (for instance, "The native says ... so and so").

An example, before concluding. Dan Sperber (1982) gives an account of a communicative sequence which is a fine example to help us to understand that the challenge of the other culture reaches the anthropologist through language, and that he cannot but treat it through language. The young anthropologist Dan Sperber finds himself in an odd situation: an Ethiopian friend, an esteemed old sage, invited him to kill the golden hearted dragon. The old man offers him a valuable piece of news ("They said to me that the golden dragon has been sighted"); and the young, strong and learned foreigner is forced to give an answer. But he cannot do other than play for time: he says: "I have no gun"; "Foreigners don't know how to find dragons", and so on; finally, he puts the answer off till the day after. Sperber underlines the actual moves of the dialogue: in fact, the two interlocutors don't oppose each other as if one believed in dragons, and the other did not. Rather, they oppose each other as an old sage who possesses a specific piece of knowledge, and a young man who, by his answers, has become a recalcitrant dragon killer. What is still more interesting is the fact that the anthropologist in his diary only recorded the first part of the conversation, that is the old man's request; whereas the anthropologist's embarrassed attempts to get himself out of the difficulty were omitted. In this way, by censoring his own behaviour, the anthropologist depicted in his diary a gap that cannot be bridged between two different cultures. Only a few years later did he fill the gap in the diary, and he tells us that he went back to that dialogue again and again. His questioning concerned the symbolic forms by means of which two cultures try to build up a communicative context, and the reasons why anthropologists are inclined to be silent about the interaction in the field as a communicative context.

By analyzing the anthropological fieldwork, I have argued that observation and interpretation in human sciences are not separable, because observation in human sciences is a form of interaction through language, and language is the symbolic structure that plays an essential role in the construction of observable objects.

Dipartimento di Filosofia
Università degli Studi di Pavia

REFERENCES

Bateson, G.: 1972, *Steps to an Ecology of Mind*, Chandler Publishing Company, New York.

Blumenberg, H.: 1960, 'Paradigmen zu einer Metaphorologie', *Archiv für Begriffgeschichte*, Vol. VI, Bouvier und Co., Bonn.

Cassirer, E.: 1923, *Philosophie der Symbolischen Formen*, Copyright Bruno Cassirer, Oxford.

Ducrot, O.: 1972, *Dire et ne pas dire. Principes de sémantique linguistique*, Hermann, Paris.

Ducrot, O.: 1973, *La preuve et le dire. Langage et logique*, avec la collaboration de M.C. Barbault et J. Depresle, Mame, Paris.

Dwyer, K.: 1982, *Moroccan Dialogues. Anthropology in Question*, J. Hopkins University Press, Baltimore.

Fabian, J.: 1971: 'Language, History and Anthropology', *Philosophy of Social Sciences*, pp. 19–47.

Favret-Saada, J.: 1977, *Les mots, la mort, les sorts: la sorcellerie dans le Bocage*, Gallimard, Paris.

Favret-Saada, J. and Contreras, J.: 1981, *Corps pour corps, enquête sur la sorcellerie dans le Bocage*, Paris.

Freud, S.: 1925. *Die Verneinung, Gesammelte Werke*, Vol. XIV, Frankfurt am Main, 1948, pp. 9–15.

Gadamer, H.G.: 1965, *Wahrheit und Methode*, Mohr, Tübingen.

Geertz, C.: 1973, *The Interpretation of Cultures*, Basic Books, New York.

Geertz, C.: 1980, *Negara: The Theatre State in Nineteenth-Century Bali*, Princeton University Press, Princeton.

Geertz, C.: 1983, *Local Knowledge. Further Essays in Interpretive Anthropology*, Basic Books, New York.

Guidieri, R.: 1980, *La route des morts*, Seuil Paris.

Hanson, N.R.: 1958, *Patterns of Discovery. An Inquiry into the Conceptual Foundations of Science*, Cambridge University Press, Cambridge.
Hesse, M.: 1966, 1970², *Models and Analogies in Science*, University of Notre Dame Press, Notre Dame (Indiana).
Holy, L. and Stuchlik, M. (eds.): 1981, *The Structure of Folk Models*, Academic Press, London.
Lévi-Strauss, C.: 1949, *Les structures élémentaires de la parenté,* Presses Universitaires de France, Paris.
Lévi-Strauss, C.: 1962, *La pensée sauvage*, Librairie Plon, Paris.
Lowry, J.: 1981, 'Theorizing "Observation"', in Pinxten, R. and Karnoouh, C. (eds.), *Observation in Anthropology, Studies in Knowledge*, Copyright Communication & Cognition, Ghent, pp. 7-23.
Malinowski, B.: 1922, 1972⁸, *Argonauts of the Western Pacific,* Routledge & Kegan Paul, London.
Malinowski, B.: 1967, *A Diary in the Strict Sense of the Term*, Routledge & Kegan Paul, London.
Meyer, M.: 1986, *De la problématologie. Philosophie, science et langage*, Mardaga, Bruxelles.
Parret, H.: 1984, 'Regularities, Rules and Strategies', *Journal of Pragmatics*, vol. 8, 4, pp. 569-592.
Pinxten, R.: 1981, 'Observation in Anthropology: Positivism and Subjectivism Combined', in Pinxten, R. and Karnoouh, C. (eds.), *Observation in Anthropology, Studies in Knowledge*, Copyright Communication & Cognition, Ghent, pp. 57-83.
Ricoeur, P.: 1970, *Freud and Philosophy*, Yale University Press, New Haven.
Ricoeur, P.: 1975, *La métaphore vive*, Seuil, Paris.
Shankman, P.: 1984, 'The Thick and the Thin: On the Interpretive Theoretical Program of Clifford Geertz', *Current Anthropology,* Vol. 25, 3, pp. 261-279.
Sperber, D.: 1982, *Le savoir des anthropologues*, Hermann, Paris.
Wagner, R.: 1975, 1981², *The Invention of Culture*, University of Chicago Press, Chicago.
Wittgenstein, L.: 1953, 1958², *Philosophical Investigations* , Blackwell, Oxford.
Wittgenstein, L.: 1958, 1969², *The Blue and the Brown Books*, Blackwell, Oxford.

VINCENZO CAPPELLETTI

ON THE DYNAMICS OF SCIENTIFIC PARADIGMS

For the historian axioms are live realities, similar to organisms which are expected to enrich their own structure, to acquire a larger dimension, to compete or coexist with different organisms. Continuing to use this metaphor - perhaps beyond the point which pleases the defenders of individuality, of the *hékaston*, in biomedicine - we could say that history requires axioms to lend themselves to fertile hybridizations. The author of some of these being the daily exchange of ideas in the scientific community, while the promoter of others is historiography, a historiography conceived as reason which enlivens, rethinks and judges the event, and as an arduous, yet necessary, drawing closer together of the absolute and the relative - the "I think" and the "what" I think - in order to convert the *esse* and the sum of *factum* and *fieri* mutually one into the other. To this end the historiography of scientific thought, and any other kind of historiographic reflection, are necessary. If the growth of the axiomatic organism can be called "dynamics", the competition and coexistence which we mentioned are rather its "dialetics". Only after this has taken place, due to history in the first place, and then to the historian, is an axiom worthy of belonging to the scientific knowledge of the world: a modern ideality born out of the ancient vocation to philosophizing and, according to the wishes of some, meant to be proposed again, unchanged, yet improved and organically related to man's position in the universe.

We are a century away from two works which were to deeply influence science and science historiography in the successive decades to come, namely: *Die Mechanik in ihrer Entwicklung*, by Ernst Mach[1], and *Einleitung in die Geisteswissenschaften* by Wilhelm Dilthey.[2] Both were published in 1883, but each one in a different cultural climate. When Mach issues his *Mechanics in its Development*, he is in Prague, in those days a creative and lively periphery of Germanic and Slav culture. Dilthey, the author of *Introduction to Humanistic Sciences*, having left the not distant Breslau, has spent one year in Berlin, the center of European science. Mach's critical attitude - Einstein calls it "Skepsis", in his *Autobiography*[3] - connoting an *inclination to doubt* rather

D. Batens and J. P. van Bendegem (eds.), Theory and Experiment, 243–258.

than *skepticism*,goes along with a certain lack of clarity and formal elegance, and an uncertain foundation of the theoretical construction. Not only does Mach ignore Hegel and refers directly to Kant - which is a characteristic of all of the Middle European thought - but he engages in a controversy with the "metaphysical" Kant and would like to return to Hume. And yet, Mach's *Mechanics* gives birth to a new historiography of science which penetrates in the movement of scientific ideas, and, being free from submission to tradition, consciously clarifies and proposes anew, without confining to mere collecting and ordering. Mach's work centers around a basic thesis: the relationship between man and nature, between the subject and the object of naturalistic cognition, is created through "Empfindung", "sensation". Sensation is "instinctive cognition", "instinktive Erkenntnis". Through sensation we come to know facts and, in a confused way, those principles which science successively isolates and defines with the help of abstraction and hypothesis. Like all empiricists, Mach is bound to educe all cognition from a single source. But what is interesting in him is not so much empiricism as the original consequences which he draws from the postulate of "Empfindung". The first one relates to historiography which becomes analysis and recomposition of traditional knowledge, starting out from very specific experience situations which can be reconstructed on the texts and attributed to various Authors. Also W. Whewell's *History of the Inductive Sciences*[4] had effectively criticized tradition, opposing two modes of knowledge: Archimedes and Aristotle, science which defines that which can be observed, and science which seeks to explain it. *Mechanics in its Development* has the importance of a manifesto, which *History of the Inductive Sciences* had not had. Both the geologist Whewell as well as the physicist Mach, are *internalists* - the distinction between *internal and external history* existed already in the Nineteenth Century, for instance in the Italian A.C. de Meis[5] - but Mach is a radical internalist and, what is more, an original one, inasmuch as in one science he seeks the traces of another, in physics he follows the path of psychology. As a historian, Mach perceives and turns into a problem something which does not stimulate the researcher's attentions. What, for instance? The following: physical cognition of the world is made up of magnitudes, measures and laws, but also of an heterogeneous element which is not formal, that is, subjective, and which is not mechanical, that is, material. The non-formal character of "Empfindung" allows the assumption of its objective reality which the non-mechanical character, on the contrary, seems to exclude. But aren't Mach's theses obvious? They are not, because of the antinomy which they contain and which we have tried to outline. An antinomy always has deep roots in the axioms of science. In Mach, who criticizes the traditional order of mechanics, a lessened tradition of the concept of nature continues to exist next to an inkling of a broadening of the horizon which ensues from the dynamics of a new paradigm: psychology. For Kant, sensibility, "Sinnlichkeit" was "Rezeptivität", receptivity:[6] not a 'doing', but an 'enduring'. With Müller's law of specific sensorial energies, sensation had again turned to be a process, i.e. original reality. And the science of sensation had become part of the group of those which deal with real processes and speak of the world. Thus, it was precisely to the concept of the world, to the heart of scientific objectivity, that the antinomy of sensation, which subtends *Mechanics in its Development*, was leading back.

In his academic lecture *Das Denken in der Medizin*, of 1887, H. Helmholtz had remembered Müller and the law of specific energies discovered by him.[8] And he had expressed an opinion which, coming from a man of such intellectual prestige, could neither be a gratuitous hyperbole nor rhetoric for the occasion. According to Helmholtz,the importance of the law discovered by Müller was equal to that of the law of gravity discovered by Newton. In the Helmholtzian lecture this statement is not explained and we must resort to suppositions. And the first and most obvious one is that Müller's discovery attributed a real, or rather, realistic basis to the physiology of the sense organs - an intermediate research program between neurophysiology and psychophysiology which Helmholtz had defined in those years in *Handbuch der physiologischen Optik*[9] and in *Die Lehre von den Tonempfindungen*[10]. It is not the science of appearance that is being pursued: having become the law of appearing, appearance transforms into reality as phenomenologists like Husserl and Heidegger, who were tied to Kant as Helmholtz had been, will observe at the onset of the new century. Helmholtz had already spoken about Müller and his discovery of specific sensorial energies on a preceding occasion, namely during his lecture *Über das Ziel und die Fortschritte der Naturwissenschaft,* which he held in 1869, at the German Naturalists' Congress.[11] Müller had done away with the theory which presupposed the "innate harmony between senses and the outside world". Sensations are not "copies", "Abbilder", but signs, "Zeichen"[12] which we learn to read inasmuch as we are involved in a process of continuous adaptation to the outside world. In short: sensation is not impression, as Kant held it, but is something specific. That which is in us succeeds in getting in touch with that which is outside of us, and, vice versa, in order to be recognized and assert itself, that which is outside of us must transform into that which is inside of us, in one of the signs which make up the sensorial universe. The theme of the passage of the *outside of us* into the *in us*, is also dealt with in other works of Helmholtz – *Über die Tatsachen, die der Geometrie zum Grunde liegen*,[13] *Die Tatsachen in der Wahrnehmung*,[14] – wherein Helmholtz searches for that which can unify the world in us and the world outside of us in one and the same universe, sustained by one single principle (the causality principle, in the first of the quoted texts). It is a subject of capital importance for scientific knowledge which will escape the sensationalist Mach. Yet, a few years later, while Helmholtz is still alive and active at the Humboldt University, Mach too promotes an originally realistic elaboration of the theme "Empfindung". In *Die Analyse der Empfindungen*,[15] which came out in Vienna in 1886, three years after *Mechanics*, Mach writes: "For us the world does not consist of mysterious essences, which, in interacting with another equally mysterious essence, the self, bring forth those "sensations" which, alone, are supposedly accessible to us. For us, colours, sounds, spaces, epochs... are the temporarily ultimate elements of which we must investigate the given connection. To fathom (die Ergründung) reality thoroughly, means precisely doing that",[16]. It would seem as if sensation had come to identify itself with that "fact in itself", "Tatsache an sich", which is represented and revealed in *Mechanics in its Development*. And we have a proof: in a note on the quoted page, Mach recalls how he reached that point in which he freed himself from the Kantian "thing-in-itself", a few years after having read Kant's *Prolegomena* and having been "powerfully, indelibly impressed" by it. One summer day, the world and the self seemed to him to consist of sensations: the self more compact, the world less. A thesis which ceases to be paradoxical if we consider it as a claim of the full reality of "Empfindung" and as an attempt to extend the domain of sensing from the I, towards nature.

The dynamics of a paradigm develops on the surface and in the depth of scientific thought: on the level of terminology and in the formal space of theories. In the historical process which was to end with the transformation of mechanics into mechanism, from Galileo to Laplace and to Helmholtz, we find both moments. Many – such as physicists, iatromechanists, chemists – speak of motions of the smallest part of matter: beside the perceptible movements there are the "occult" ones, the "verborgene Bewegungen", to use the Helmholtzian terminology. But in the dynamics of the mechanic paradigm and its dialectics with other paradigms, that which matters is the structural moment: the attempts to justify the priority of the science of movement, the challenge of the other axioms, the correlation of the synthesis. Helmholtz's early memoir

Über die Erhaltung der Kraft[17] had inferred the principle of the conservation of "force", then energy, from the universality of movement and from the constancy of the sum of "tension forces" and "live forces", i.e. of potential energy and kinetic energy. In dynamics and in the dialectics of mechanics and mechanism, force and structure concepts are fundamental. The "mystère des forces", which Euler frankly records in his *Lettres à une princesse d'Allemagne*,[18] the "indestructible qualities", "unvertilgbare Qualitäten", which Helmholtz postulated in order to give a meaning to the possibility itself of an energy conservation problem in the passage of natural processes from one class to another; the "Bau", the "structure", the "group" - "Gruppe", a dense and significative notion shared by morphology and mathematics after Cuvier and Galois: all these points - limits of the formal space of mechanism were, and were going to become, the origin of opposite trajectories, of eidetic paradigms towards the science of movement. In saying "paradigm", we mean - but it is implicit in what we have said - the constitutive axiom of a scientific discipline in its present state and its virtual developments. This is the meaning which the term has come to have in R.K. Merton's *Paradigm for the Sociology of Knowledge*, an article of 1945, re-published in *The Sociology of Knowledge*, in 1973.[19] After Durkheim and Scheler, Mannheim and Sorokin, Merton has analyzed the influence of social factors on the development of science, ranging from causal motivations to functional attributions. In Merton there is a paradigmatic intersection of sociology and epistemology, as there was one in Mach of psychology and mechanics. What matters is always the structural insertion point: where, how and where a conceptual space passes through another one seeking to blend with it into complete cognition of one class of entities. Merton does not identify, but correlates the space of the social element with the epistemic one. The first space is the seat of the function of promotion, the second one that of the function, or creative condition, of knowledge. And the intersection point has, as coordinates in the two spaces, a given possibility of development of cognition, and a given offer of theoretic material. Now, if we try to define where mechanics and psychology intersect, in Mach, we reach a point: "sensation", "Empfindung", which, when more carefully observed, becomes an intensely germinative and productive area. In *Mechanics in its Development*, sensation, as we already mentioned, is a depository of immediate experience which nurtures theory through the rational process of analysis, abstraction and definition. Nature's properties cannot be invented as Mach is to write in the seventh edition of the work, in an argument with G. Vailati.[20] In this implication, sensing is the same one which

Aristotle defines in his third book of *De anima*, as two which become one, as a unity of "sensing capacity" and "sensed object", "aistheticôn" and "aisthetôn" in the sensing subject.[21] And the unification of that which is sensed and that which is intelligible, seems to come up again in the subsequent, rational, phase of scientific knowledge, even if with great uncertainty: as we have said, the Machian sensation contains the rational virtuality of the principle which explains it. The anti-metaphysical Mach gets closer to a gnoseology of realistic imprint, which - in the Vienna of his days - expressed itself through F. Brentano, Aristotle commentator and upholder of the intentional reference to the object as a distinctive characteristic of the psychic act.[22] But with the *analysis of sensations*, and with the passage of "Empfindung" from cognition to world, the dialectics of mechanics and psychology transfers its own ferment into the object of scientific cognition and becomes stimulus to re-think the essence of nature, while mechanism declines.

With *Mechanics in its Development*, the psychological paradigm has established a firm relation with the problem of scientific cognition in its two aspects: subjective and objective. Objectively, sensation appears like a still unresolved plexus of secondary and primary qualities, nor could it be otherwise, after Müller and his Newtonian discovery of specific sensorial energies. Subjectively, "Empfindung" acts as a bridge between natural science and human sciences in the awakening of the renewed epistemological interest prompted by those scholars - A. Comte, J. Stuart Mill - who had noticed a substantial enrichment and change in the objects and methods of scientific knowledge. But, in Mach, the dynamics of the psychological paradigm is accentuated in the works of the Vienna years, the *Analysis of Sensation* and *Cognition and Error*.[23] Sensations become the "temporarily ultimate elements", "vorläufige letzte Elemente",of the world.[24] Besides, the identity, or at least the convergence, of psyche and nature seems to erase the traditional criteria of distinction between objective and subjective, primary and secondary, allowing only one point of reference to subsist: the world as horizon of all that is real. According to Mach: *psychical*, means sensation in itself, *physical*, sensation analyzed and correlated to real elements. But "Empfindung", however, appears to be totally different in the new, cosmological, context as compared to the gnoseological context of Mechanics. And Mach confirms it in one of his enlightning confessions. Having published *Mechanics in its Development*, he had set out towards a "physico-psychological monadology", only to immediately change road. A confession, only partially sincere.[25] If we look through *Cognition and Error* we find in it a scenario of the psyche which extends to animals, to the primitives and to the unconscious. And meanwhile, in man the vestibular apparatus of the inner ear, the organ of static and kinetic balance perception had been analyzed. In one of the great, and little known, works of those years, *Das Ohr-Labyrinth* by E. von Cyon,[26] the vestibular apparatus is seen as an actual organ of space and time sensations. Sensing, as an objective fact, is closely correlated to the anatomic-histologic structure of the sense organs, as a subjective fact it is seen in works such as we have quoted. Therefore a distinction shows between *sensing* and *perceiving* which will be confirmed by neuro-physiological and psychological research: sensing is qualitative, in Müller's meaning, the other one is gestaltic in Ch. Ehrenfels's, and then in M. Wertheimer's meaning. Since Cuvier's paper: *Conjecture sur le sizieme sens qu'on a cru remarquer chez les chauves-souris*, much ground had been covered in analysing sensorial manifestations, even the subtle and complex ones. Behind the monadologic adventure of Mach, there was also ancient and modern prehistory of the psychological paradigm. According to Aristotle,[27] the properties of the soul are twofold: local movement and thought. In *Ethics*, Spinoza had stated that "Individua (...) quamvis diversis gradibus, animata tamen sunt",[28] because the relation between the body and an idea is a universal property of that which exists. And the Leibniz of the *Monadology*, a source of patent and occult influences on the science of the Germanic area, had detected in the "vis rapraesentativa" the essence of the monad, the cornerstone of the universe.[29] The historiography of science such as Mach had seen it, namely an analysis and reconstruction of traditional knowledge, cannot and must not ignore these antecedent facts. But in order to seek and utilize them, historiography, or better still, the historian, must accustom itself, and himself, to doubting. The Machian monadologic adventure was, perhaps, not accidental and not even ephemeral. Mach must have sensed that a construction of the universe which started out from psychological magnitudes was an extreme hypothesis, but not a paradoxical one, and even less an absurd one, after psychology had again come to be part of the sciences which deal with real processes.

From Mach to Dilthey: equally a central figure of scientific awareness and of interdisciplinary dynamics to which we feel the urge to relate. The Dilthey of *Introduction* is faced with a vast and articulated group of disciplines: psychology, anthropology, sociology, historiography. They are the "sciences of the spirit", which we prefer to call *human sciences*, given the polysemy of the term "Geist", which has turned to actual ambiguity. In the horizon which Dilthey takes into consideration, psychology and anthropology have priority: they deal with immediate experience, with perceiving and doing, which are closely connected to one another. The interaction of paradigms, which we find in Mach, is substituted, in Dilthey, by the awareness of a deep fracture inside scientific thought. On the one side, natural sciences with their long tradition, on the other, the recent sciences of man and of society. As compared to the psychophysical situation which Mach considers, the field of experience examined by Dilthey contains a different element: active and creative. Sensation specifically elaborates a stimulus which it is bound to receive from the outside: but will, expression, creativeness escape to a great extent, or totally, the schema of the relation stimulus-answer. The facts of the spirit are the upper limit of the facts of nature, says Dilthey[30] drawing spirit and nature together again, perhaps with a lapse which is revealing of an acute desire for unity behind the awareness of the epistemological fracture which we have mentioned. Besides, psychophysics does not constitute all of psychology. Psychology is the "first and most elementary science of the spirit",[31] and contains *in nuce* the features and peculiarities of the entire edifice of the "Geisteswissenschaften". Which are these features? The individual, the quality, the cognition from the inside and, as synthesis of it all, the cognition as "understanding", "Verstehen". To consider the effect of Dilthey's *Introduction* a historic turning point, is no overstatement: but the space within which the turning point proper took place, is *metahistory*, or the historiography of science. Mach's *Mechanics in its Development* and the *Introduction to Human Sciences*, both published in the same year, have something much more substantial in common than a mere temporal coincidence. Together they reach an apperception which, as we said, becomes a manifesto wherein the fertility of historiographic research on scientific traditions is not recommended but concretely proved. Fertility : but for whom and to what purpose? Fertility for science: there lies the paradoxical aspect of the Machian and Diltheyan manifesto. But fertility for a science which sets itself the aim of understanding the world from within the specialistic axioms and from the sectorial paradigms without preconceived exclusions. The world is both exterior and interior to the human subject: the exterior world is a totality of points, the interior one, a totality of individual experiences. The objective conditions postulated by natural sciences - space and movement - have no value for the sciences of the spirit for which, as Dilthey states in his *Introduction*:[32] the "sign of something real" is sufficient. And at another point he adds that their object "consists of given, not of inferred, unities, which are understandable to us from within".[33] We are witnessing the return on the scientific scene of entities which had been banned at the beginning of the modern era by the dominating mechanical paradigm: among these, quality and individual. If we read the *Introduction* of 1883, keeping in mind the author's early works, we understand the scope and meaning of Dilthey's work even better. "The secret of the world" - he wrote between 1874 and 1879 - "is individuality indivisible, indissoluble".[34] And he added that the psyche of the universe is an immanent finality resulting from the forces of the minimal parts. And he ended with a courageous and classically clear aphorism: "The soul of the universe is an abyss of individuality", because "Geist" is synthesis and entelechy.[35] There is allusiveness in all this, and dense shadows bereave the light of its space. But there is also a promise of clarity beyond the limits of the present. Historiography knows the deceptions and the fecundity which are comprised and interwoven in each one of man's days and it can open science's eyes to fundamental intuitions.

In Mach, sensations are "elements", and in Dilthey they are "facts of the spirit", one of the three classes of spiritual facts, together with "behaviors" and "rules". But Dilthey does not confine himself to amplifying the prospective relating to the constitution of the non-mechanical half of the world. He seeks their particular characteristics and speaks of "quality", "individuality", of facts which are "comprehended" inasmuch as they are known from within. We have said that in Mach the intersection of mechanics and psychology takes place in an area where the borders are vague: however, in the center of this area we can attempt to place the qualitative fact or moment. The dense, unresolved contents which sensations bring into the theory of movement and, in general, into scientific theory, are undoubtedly qualities or groups of qualities. In Dilthey, to the contrary, quality is that which creates a radical difference between natural science and humanistic sciences, generalizing knowledge and individualizing knowledge, quantitative knowledge and, precisely, qualitative knowledge. The defeat of metaphysics "queen of a kingdom of shadows"[36] stems from quality, from the already mentioned abyss of the individual which quality reveals at each step of the structure of the world. And the question is this: does Dilthey perceive the dynamics of the paradigm which he has instituted and which contains also psychology, interacting in Mach with mechanics, or does he not? Does he promote it, or does he not? The greater part of Dilthey's *Introduction* is undoubtedly taken up by the definition of the constitutive axioms of the sciences of the spirit - the humanistic sciences, as we say. But the work itself is filled with observations oriented in the opposite direction, namely towards the consciousness of "everything from nature"[37] towards handing back to metaphysics the "possible task (...) of enclosing the results of positive sciences in a general vision of the world",[38] and towards the acknowledgment that man's spiritual life is only one part of his live, psychophysical wholeness[39]. And the *Introduction* is permeated by references to a "gnoseologic foundation of the sciences of the spirit"[40] subsequently to the fact that "a shortsighted specialistic activity" has divided science from philosophy and prevented a logic or a methodology to take shape which would create the internal unity of science and replace, thus, the old metaphysical connection of cognitions[41]. But in this case, our historiographic reading of Mach and Dilthey must go beyond the texts and pin-point present requirements in both, in order to outline a dynamic synthesis of paradigms. We repeat it: in Mach, there is a graft of psychology on mechanics, in Dilthey there is a detection of a fracture in scientific knowledge and the urge of a unification on new bases. Can we say that Mach is close to Dilthey, or that Dilthey is close to Mach? Yes, we can, with a rational experiment which transfers Diltey's notion of individuality-quality into the conceptual texture of *Mechanics in its Development*. Sensation, such as Mach defines it, is undoubtedly individual-qualitative. That which, according to Dilthey, is a fact of the spirit, in Mach's *Mechanics* is a moment of the theory of a class of physical facts, the motion phenomena. Could the motion phenomena be lacking in individuality-quality, while the sensations inherent in the theory of these phenomena are endowed with it? No, this is not possible: there would be two worlds and not one, two sciences and not one, split by an incurable fracture. And the historical achievement of science would then have amounted to the absurd resolve of human language speaking about that which is totally other than man. Dilthey says: space and movement have no value for the sciences of the spirit.[42] This is more than we can grant him. As long as the "Geist" comes to be as a "Tatsache", said "Geist" is inside time and space and moves. Undoubtedly, given equal conditions internal to the process from which the spiritual fact is born, the latter can show at any given point of isotropic space-time. But, we repeat: the spiritual fact is also a space-time fact and this could not happen if spirit and nature were not different modes of one and the same reality, orders which correspond to one another according to an essential analogy. Transposed in Mach, Dilthey clarifies the individual-qualitative essence of "Empfindung", of the Machian "sensation", but the latter leads back Dilthey's "fact of the spirit" to a totality, to a horizon which includes also nature.

Towards the end of the century, a three-dimensional *sensorial vector* constitutes itself inside the psychological paradigm, namely: Mach's element-sensation, Helmholtz's sign-sensation and Dilthey's individual-sensation (strictly speaking, 'perception'). Mach's and Dilthey's stances are, or seem to be, at the extreme opposites. Helmholtz is at the center, in the space of traditional neurophysiology, anchored as he is to the concept of "sense organ", but with a substantial graft of physical methodology in optics and physiological acoustics. But all three: Mach, Helmholtz and Dilthey, are subject to the consequences of group structure which the sensorial vector has taken on. In Mach, initially, the importance and function of sensation and sensing are gnoseologic, but subsequently the latter become accessible to that Spinozian and Leibnizian meaning of sensing which we have considered to be part of the presuppositions to the 19th Century developments of psychology. Helmholtz fully understands the innovative scope of Müller's law of specific sensorial energies. It can be said that it is precisely with him that the concept of a copy-sensation ends, and an ancient, Aristotelian and Plotinian meaning of sensing as form reappears: a sensorial form which is connected to physical form, other and different. And, at last, Dilthey: according to him, sensing is "spiritual" and not "natural", however the problem of nature re-emerges in the Diltheyian page, with anti-mechanistic traces, while a conception of the universe as a system of individuals-quality outlines, and the theme of finality, immanent in the individual, diffuses Aristotelian moods in the interstices of the *Introduction*. The ambivalence of the basic psychologic vastity favours an ample circulation of the psychic element in the energetic and biological one. Not all of that which is formal, eidetic, structural and qualitative in the science of the early 20th Century stems from the dynamics of psychology: the influence of mathematical formalism and geometric rationalism is strong and decisive. However, basically, one type of nature which was occupied by mechanics, de-essentialized and materialized, is gradually set against another type of nature which is inhabited by something which it is outmoded to call form, risky to define as spirit and impossible to characterize as psyche, because psyche is sensation in the early Mach and in Fechner, specific sensing in Müller and Helmholtz, perceptive field in Ehrenfels and Wertheimer and, in Wundt it is the series of psychic elements listed by himself. Spinoza's "individua omnia animata" and Leibniz's "vis rapraesentativa" are inaccessible: a patent lack of determinations and mediations severs them from the dominating axioms in the history of scientific thought. Quality, to the contrary, is meant to cover a great deal of ground in the wake of concepts such as organized complexity-structure-information. The embryologist H.

Driesch is wrong in not leaving the factor (E, from entelécheia) of development regulation in tunicates, and other morphophysiological processes, in a "parking" area, an area of decantation, and to transform it into a "psychoid", tying it - even though very cautiously and with many limitations - to a paradigm which cannot receive it[43] Some deem it necessary to re-examine fundamental traditions and conventions, as Mach had done in mechanics: namely the "energetists" and specifically the chemico-physicists who gather around the "Zeitschrift für physikalische Chemie". Their spokesman is W. Ostwald, and in him the historian, who is unaware of the tight web of his relations with the Author of *Mechanics in its Development*, could detect unusual cues, such as: the dedication of his major work to Mach,[44] the intention of starting an historically based "philosophy of nature"[45] the attempt of reabsorbing the "Geist" in science's conceptual horizon, substituting "force" with "energy" and overcoming the mechanistic preliminary issue, derived from Helmholtz, but actually Kantian.

Behind Ostwald and the energetists - an intellectual minority which is of particular interest to the historian - we find an intuition and an illusion. The intuition perceives the fracture which has come about in the axioms of science during the course of he mechanical age. Thought and extension, force and matter, "Geist" and "Natur" have formed couples of antithetical terms - antithetical and unrelated. In order to formulate the principle of the conservation of force, according to Ostwald, Helmholtz has had to reduce all forces to mechanical work[46]. This is not quite so. In Helmholtz the "indestructible qualities" had remained, pertaining to the various classes of phenomena - electric, magnetic, chemical, thermodynamic - but acting as "tension forces" and "live" forces, with a passage which the reader perceives even today, from the mechanical to the eidetic moment. Reopening a long controversy, Ostwald opposes J.R. Mayer, the first advocate of energy conservation, to Helmholtz. With his essay of 1842, *Uber die Kräfte der unbelebten Natur*[47], Mayer recognized that forces are qualitative, pre-mechanical entities, thus opening them to a non-mechanical unification of scientific axioms, into which also mechanics could then be enclosed. In the background of "Kraft", there is also "Geist", with all the richness and vagueness of its connotation. But it is a relative vagueness: "Geist" is also meant as "psychische Erscheinungen", "psychic manifestations", and Ostwald writes: "I deem it possible to subordinate all manifestations of the psyche to the concept of energy".[48] It is a pity that Ostwald dedicated his *Naturphilosophie* to Mach and not also to Dilthey: from an eminent science historian such as Ostwald, we could have expected the meaningful comparison between the Authors of *Mechanics* and of *Introduction*. The Prague and Berlin dialectics could have been expressed in Leipzig, Ostwald's university and one of the major academic seats of psychological research. But the fact that no encounter between Ostwald and Dilthey took place, is not haphazard. Energetics remained a phenomenologic theory which was unable to rise to axiomatics; and phenomenology orders experience temporarily, but does not know how to interpret it, as it cannot define it. In Ostwald and in the Mayer whom the former praises, it would be vain to look for pages corresponding to the crystal clear "introduction" to Helmholtz's essay *On the Conservation of Force*. But it would be equally vain to look in Helmholtz for the perception of that particular tension between the non-ponderable, eidetic moment and the ponderable, mechanic one, which had been clear-cut in Mayer and which is sharp to the utmost in Ostwald. As for Dilthey, in postulating the necessity of a gnoseology of the sciences of the spirit, he had clearly shown the borderline between phenomenology and institution, or re-institution, of the "Geist" in science. The dynamics of the psychological paradigm, the movement of its ambiguous but meaningful sensorial vector pass - after Mach and Dilthey - at any rate through a minority ideology: energetics. Energetics and psychobiology, the Drieschian and, above all, the Neo-Lamarckian one, the latter one mediated through the psychiatrist E. Bleuler,[49], converge, in turn, in the psychological field, in the work of C.G. Jung, Bleuler's pupil in Zürich. The passage of the "libido" from a pulsional force, as it was in Freud, to "psychic energy", as it is in Jung, would be a mere question of terminology, if energy were not, in Jung, as in the energetists, a qualitative vector and an evolutive one. In a paper of the late Jung, *Über die Energetik der Seele*,[50] energetics is seen to have risen to the full consciousness of being a qualitative, structural, finalistic theory of becoming. Energy, according to Jung, is a plexus of relations which can limit and specify itself as a mechanical force, but which can penetrate into the "psychological machine" of the symbol in order to have the synthesis of subconscious and consciousness come out of it. We are still on phenomenological grounds with a deep meaning of the "Geist" which appears, here, as "Seele", soul, which, in turn, becomes psyche again, "psychic energy", in the English edition of Jung's works. The polisemy of single terms and the synonymy relating to single concepts which attempt to cross the borders of paradigms, are phenomena which do not surprise the historian. To the contrary, for him they represent signs of rational vitality, evidences of intellectual demands which are not satisfied by the historical set up of knowledge; they mean the premises and the promises of comparison and interactions between theories, paradigms and axioms.

The dynamics of paradigms specifies the latter's origin, its formal developments and the mutual interactions. Its presupposition is the unity of nature. Nature *must be* constituted by classes of beings which are structurally related to one another, in order that a paradigm which has a class of beings as its formal object, *may be able* to interact with a different paradigm. Also the meta-unity of scientific knowledge and of science historiography depends on the same presupposition. Particularly this latter one, namely science historiography, sinks to mere traditional or academic nomenclature, if it does not assume in its own object the formal movements and changes of the axioms in their mutual relations: that which we have called - here as elsewhere - the dynamics and dialectics of scientific paradigms.

University of Rome

NOTES

1. E. MACH, *Die Mechanik in ihrer Entwickelung historisch-kritisch dargestellt,* Leipzig 1883, 1912[7], posthumous editions 1921[8] and 1933[9] by L. Mach.
2. W. DILTHEY, Einleitung in die Geisteswissenschaften. Versuch einer Grundlegung für das Studium der Gesellschaft und der Gechichte, I, Leipzig 1883, then in *Gesammelte Schriften,* I, Leipzig-Berlin 1923 and Stuttgart-Göttingen 1959 and 1966[4].
3. A. EINSTEIN, Autobiographical notes,in *Albert Einstein philosopher-scientist*, ed. by P.A. Schilpp, New York 1951[2], p.20.
4. W.WHEWELL, *History of the inductive sciences from the earliest to the present time*, I-III, London 1837, and I-IV, London 1847[2], and I-III London 1857[3].
5. A.C. DE MEIS, *Delle prime linee della patologia storica*. Prefazione al corso di storia della medicina, Bologna 1866, p.8.
6. I. KANT, *Kritik der reinen Vernunft* (1787 in *Werke*, hrsg. von E.Cassirer, III, Berlin 1913, p.55.
7. J. MÜLLER, *Über die phantastischen Gesichtserscheinungen.* Eine physiologische Untersuchung mit einer physiologischen Urkunde des Aristoteles über den Traum, Koblenz-Bonn 1826.
8. H. HELMHOLTZ, *Das Denken in der Medicin*. Rede gehalten zur Feier der Militairärztlichen Bildungs Anstalten in Berlin 1877, in *Vorträge und Reden*, II, Braunschweig 1903[5], pp.181-182.

9. H. HELMHOLTZ, *Handbuch der physiologischen Optik*, 1-3, Leipzig 1856-'67, Hamburg 1885-'86[2].
10. H. HELMHOLTZ, *Die Lehre von der Tonempfindungen als physiologische Grundlage für die Theorie der Musik*, Braunschweig 1863, ivi 1877[4].
11. H. HELMHOLTZ, *Über das Ziel und die Fortschritte der Naturwissenschaft*. Eröffnungsrede für die Naturforscherversammlung zu Innsbruck (1869), in *Vorträge* etc., I, Leipzig 1903[6], pp.367-398.
12. Op. cit., p.393.
13. H. HELMHOLTZ, *Über die Tatsachen, die der Geometrie zum Grunde liegen,* 1868 in *Wissenschaftliche Abhandlungen*, II, Leipzig 1883.
14. H. HELMHOLTZ, *Die Tatsachen in der Wahrnehmung*. Rede gehalten zur Stiftungsfeier der Friedrich-Wilhelms-Universität zu Berlin, 1878, in *Vorträge* etc., cit., II, Braunschweig 1903[5], pp.213-247.
15. E. MACH, *Beiträge zur Analyse der Empfindungen*, Jena 1886, then published with the title *Die Analyse der Empfindungen und das Verhältnis des Physischen zum Psychischen*, Jena 1900[2], 1922[5].
16. E. MACH, *Die Analyse* etc., cit., Jena 1919[8], pp.24-25.
17. H. HELMHOLTZ, *Über die Erhaltung der Kraft*, eine physikalische Abhandlung vorgetragen in der Sitzung der physikalischen Gesellschaft zu Berlin am 23 Juli 1847, Berlin 1847.
18. L. EULER, *Lettres à une princesse d'Allemagne* sur divers sujets de physique et de phylosophie, I-III, St. Petersbourg 1768-'72, Mietan-Leipzig 1770-'74[2].
19. R. K. MERTON, *Paradigm for the sociology of knowledge*, in *The sociology of science*. Theoretical and empirical investigations, Chicago-London 1973, 1974[2], pp.7-40. The original title of the article is *Sociology of knowledge*, in *Twentieth-Century Sociology*, by G. Gurvitch and W.E. Moore, New York 1945, pp.366-405.
20. E. MACH, *Die Mechanik* etc., cit., p.47.
21. *De anima*, 426 a 11.
22. F. BRENTANO, *Psychologie vom empirischen Standpunkt*, Wien 1874, Leipzig 1924[2].
23. E. MACH, *Erkenntnis und Irrtum*. Skizzen zur Psychologie der Forschung, Leipzig 1905.
24. E. MACH, *Die Analyse* etc., cit., p.24.
25. Op. cit., p.24 n.l.
26. E. von CYON *Das Ohrlabyrinth als Organ der mathematischen Sinne für Raum und Zeit*, Berlin 1908.
27. *De Anima* 427 a 17-20.

28. *Ethices* pars II, propositio XIII scholium.
29. *G.G. Leibnitii Opera philosophica omnia, inst. J.E. Erdmann*, Berolini MDCCCXL, pp.705-712.
30. W. DILTHEY, *Einleitung* etc., in *Gesammelte Schriften*, I, Leipzig-Berlin 1923, p.17.
31. Op. cit., p.33.
32. Op. cit., p.119.
33. Op. cit., p.109.
34. W. DILTHEY, *Erkenntnistheoretische Fragmente (1874-1879)*, 1.Übersicht der Möglichkeit einer konsequenten Gestaltung des Empirismus, durch welche die Einsicht in die Objektivität der Erscheinungen begründet würde, in *Gesammelte Schriften*, Göttingen, 1977, XVIII, p.197.
35. Op. cit., p.198.
36. W. DILTHEY, *Einleitung* etc., cit., p.405.
37. Op. cit., p.17.
38. Op. cit., p.358.
39. Op. cit., p.15.
40. Op. cit., p.116.
41. Op. cit., p.271 s.
42. Op. cit., p.119.
43. H. DRIESCH, *Philosophie des Organischen*, Leipzig 1928^4, pp.109 ss., 259 ss.
44. W. OSTWALD, *Vorlesungen über Natur philosophie*, Leipzig 1902 (Ernst Mach gewidmet).
45. Op. cit., pp.3 ss.
46. W. OSTWALD, *Die Überwindung des wissenschaftlichen Materialismus*, in *Abhandlungen und Vorträge allgemeinen Inhaltes*, Leipzig 1904, pp.232 s.
47. R. MAYER, *Bemerkungen über die Kräfte der unbelebten Natur* 1842), in *Die Mechanik der Wärme*, Stuttgart 1893^3, pp.23-32.
48. W. OSTWALD, *Die philosophische Bedeutung der Energetik* (1903), in *Abhandlungen* etc., cit., p.278.
49. E. BLEULER, *Naturgeschichte der Seele und ihres Bewusstwerdens*. Eine Elementarpsychologie, Berlin 1921; *Die Psychoide als Prinzip der organischen Entwicklung*, Berlin 1925.
50. C.G. JUNG, *Über die Energetik der Seele* (1948), in *Gesammelte Werke*, Olten und Freiburg im Breisgau, VIII, 1971, pp.1-74 (*On psychic energy*, in *Collected Works*, VIII, New York 1960, pp.3-66).

MARCELLO PERA

BREAKING THE LINK BETWEEN METHODOLOGY AND RATIONALITY. A PLEA FOR RHETORIC IN SCIENTIFIC INQUIRY

1. INTRODUCTION

"What ultimately decides the fate of a theory is the result of a test".[1] I do not know of any past or present-day scientists who would not readily agree with this favourite maxim of Popper's. It has much to be said for it, indeed: as a description of what is going on in science, it seems to be adequate; as advice, it is easy to apprehend and, at first sight, easy to apply; as a slogan for publicity, it is appealing. It voices the old, venerated tradition that science is an objective, impersonal game whose only players are ideas and facts, without any intrusion of subjective factors unless they can be isolated or rendered harmless. Science, Popper also says, is "knowledge without a knowing subject";[2] likewise, Galileo remarked that science only relies on "sensible experience and necessary demonstrations", so that "in the natural science the art of oratory is ineffective".[3]

But if we take this view seriously, we are faced with the problem of explaining how scientists happen to disagree and, what is more paradoxical from this point of view, why there are so many cases in which they continue to disagree even when they recognize the same facts as being pertinent to their ideas. What solutions may be adopted to explain this fact? Three seems to be viable.

First, we might suppose that scientists are hypocritical and preach what they do not practice, in the sense that when they carry on their work, either they follow different rules or they obey no rules at all. Incidentally, this might explain why some of them, Einstein for example, spoke of scientists as opportunists and advised us to look at what they actually do instead of what they claim they do. Second, we might suppose that disagreement in science depends on personal idiosyncrasies which may be if not completely eliminated, at least considerably mitigated by binding scientific inquiry by more sophisticated methodological rules. Third, we might take the view that methodology does not totally regiment scientific decisions and that science is less of an impersonal enterprise than scientists would have us believe.

In different ways and to different degrees, all three solutions

D. Batens and J. P. van Bendegem (eds.), Theory and Experiment, 259–276.

I have just mentioned have been recently advocated by someone or other. In particular, Feyerabend upholds the first, contributing, as nobody else has, to shaking our faith in methodology; hoping to save as much as possible of the lost paradise, Lakatos and his school have stuck to the second and tried to sophisticate the received methodological rules; Kuhn proposes the third solution.

In this paper I shall side with Kuhn, but I wish to add further reasons for advocating that the rationality of science does not depend on methodology. In particular, I shall be focusing on two points, namely: (*i*) can methodology completely cover or regiment scientific decisions? (*ii*) if not, what should we add to, or put in the place of, methodology in order to keep scientific decisions rational?

To state my point of view right away, my answer to question (*i*) will be that, owing to its inevitable imprecision, vagueness, incompleteness, no set of methodological rules, however elaborate, or no scientific code, however detailed, can ever cope with the actual phenomenology of research. Even the best methodology presents gaps which cannot be filled unless personal factors are introduced.

Since, as I shall try to argue, this limitation is constitutive of any methodology, my proposal for question (*ii*) will be that scientific rationality does not depend on the kind of general methodological rules scientists may refer to, but rather on the kind of specific arguments they actually make use of to take their decisions. To capture the multiform variety of these arguments, standard deductive and inductive logic is not enough. This holds good not only in the most relevant cases of deep theory-change, as Kuhn has convincingly argued, but even in the most elementary cases of scientific routine, as I shall try to show. If this is the case, my suggestion is that we should enlarge our methods of analysis and take into consideration even those kinds of arguments which, in spite of their being neither formal demonstrations nor empirical proofs, are nonetheless rational ways of reasoning. Thus, a rhetoric of scientific investigation is my answer to the limitations of methodology. If this rhetoric has not yet been examined, or, when it has been envisaged, it has been assimilated to propaganda or irrational means,[4] it is because the problem of methodology continues to be ill-conceived. Ever since Descartes, we have tended to identify what is scientific with what is rational and what is rational with what may be submitted to strict, clear-cut rules of method. A "Cartesian syndrome", as it were, still affects many philosophers' minds.

I shall divide this paper into three parts. First, I shall focus on the intrinsic limitations of any methodology; second, I shall suggest that only rhetorical arguments can overcome such limitations; finally, I shall make use of a historical example to show that scientists do resort to rhetoric in taking their decisions. Since this paper is part of a larger project, some relevant questions will be here deliberately left aside.[5]

2. THE LIMITATIONS OF METHODOLOGY

If one finds that scientific research, as actually practiced by working scientists and witnessed by the history of science, does not comply with the received methodological rules, and, most of all, if, under the effects of the "Cartesian syndrome", one presumes that the only alternative to rule-dominated or standard-governed research is "mob psychology", then the most apparently obvious solution he may resort to is to try to improve those rules by rendering them more adequate.

This seems to have been the normal reaction to Kuhn's challenge and this, in any case, is the core of the sophisticationist programme as conceived by Lakatos and carried on by him and his school. According to this programme, the aim of methodology is to find definitions of science or demarcation criteria or criteria of theory-preference which are at the same time universal and able to capture scientific practice, or at least the best exemplars of it.[6]

I take this programme as "degenerating" according to its own standards. In my view, sophisticationists have reacted to the difficulties and crises of methodology in the same way as Ptolemaics reacted to the obstinate resistance of facts to be projected onto the screen of the circles and epicycles they had devised: as the reconstruction of episodes of the history of science went on, and as the gap between such episodes and the received methodological rules proved to be increasing, sophisticationists have complicated these rules in order to make them suitable for the new fields. The outcome has been a *post factum* and *ad hoc* methodology with two main features: it is universal at the cost of being generic; and it is historically adequate at the cost of being void of any heuristic content.

Elsewhere I have examined what I consider the main shortcomings of this programme.[7] In principle, I do not believe that the history of science can be used as a test of methodology; and, as a matter of fact, I do not believe that, as a result of their tests, sophisticationists did succeed in expurgating scientific decisions of all personal factors and in submitting them to objective, impersonal rules of method. Thus, to the extent that the programme maintains that scientific rationality depends on scientists respecting universal and impersonal standards, it fails. Here I intend to press my argument a step farther; I intend to show that a fairly general reason exists which should alert us to the fact that the sophisticationist programme cannot attain its goal and which might suggest us where we may find a better alternative.

The reason I allude to is that all methodologies have their limitations. This has to be taken not only in the sense Feyerabend made us familiar with, that is to say that scientists feel themselves free to violate well-accepted rules, to suspend them or to invent new ones, but primarily in the sense that any methodology, however accurate and detailed, possesses an "open texture". The limitations spring first and foremost from the *text* of the rules, from the inside, as it were.

An analogy may help to clarify the situation as I see it. As far as scientific practice is concerned, a methodology is like a code of political, social and civil relationships; and a scientist's attitude towards a scientific code is the same as a judge's towards a legal code. When a scientist takes a decision – for example, when he decides that such and such a theory is acceptable or is to be rejected – he is in the same position as a judge who brings in a verdict – for example, when he declares that a certain kind of behaviour is permitted or is to be condemned. The positive as well as the negative analogies between the judge and the scientist may shed light upon the structural limitations of methodology I have in mind.

Let us take a look at the positive analogies first. Both the scientist and the judge have to take certain steps before they can arrive at their final resolutions. In particular, three steps seem to be a pre-requisite of any theoretical appraisal of cognitive claims and of any practical decision concerning our attitude towards cognitive claims. A scientist must:

– *ascertain* a logical and factual situation. For example, is the observational result or report O reliable? Does it deductively follow from the given theory T or is it simply suggested by T? Does T contradict T'? Is T compatible with other theories which have currency in established knowledge? In addition to that, a scientist must:

– *choose* the pertinent norms under which to subsume the logical and factual situation already stated. For example, if O contradicts T and T seems to have no valid alternative (or successfully explains other relevant facts or gains considerable support from other important theories, etc.), should we consider T as being refuted by O and then reject it or should we resort to more tolerant norms? Finally, once the situation is ascertained and the

pertinent norms chosen, a scientist must:
- *interpret* the prescriptive content of the norms and establish the limits of their application. For example, if the scientific code in use contains the norm "Do not introduce *ad hoc* hypotheses", should we take this to mean that such a norm holds good for all *ad hoc* hypotheses, irrespective of the merits of the theory they are called upon to save and of the consequences a literal application of the norm may produce, or should we take these circumstances into consideration? And what does an "*ad hoc* hypothesis" mean?

Even a quick look at these questions is enough to show that none of the preliminary decisions a scientist is called upon to make before taking a final resolution concerning a cognitive claim, can be assumed without introducing personal factors. The point is that methodological codes, just like legal codes, do not possess any of the properties of a formal system or of a game. The rules of the "game of science", to use an expression of Popper's, are less transparent and less stringent than the rules of football or chess.

First, the prescriptive content of methodological rules can hardly be said to be precise. Even an apparently clear rule like "Reject hypotheses contradicted by well-supported observational results" is ambiguous, for "well-supported" may be understood differently by different people. We could, of course, combine the rule with proper definitions, but there is no way to eliminate all ambiguities and imprecisions without rendering the whole code very complicated, artificial, hardly manageable and sooner or later inadequate for scientific practice.

Second, just like a legal code, and unlike a formal system or a game, a methodological code may lead to antinomies: there may be situations in which a rule of the code induces one decision while another rule induces or recommends a different one. Consider, for example, a methodological code which contains the following rule of rejection:

R_r "Reject any hypotheses whose observational consequences are contradicted by observational data which do not constitute a local or minor anomaly".

A rule like this may well appear in a Popperian code, for, as Popper remarked even before Lakatos, in science we must leave room for some dogmatism and allow any theory to fully develop its potentialities before rejecting it as a result of evidence to the contrary. But a Popperian code also contains the following rule of acceptance:

R_a "Accept only those hypotheses which are severely tested".

It should be noted that, in Popper's interpretation, this rule excludes the acceptance of those hypotheses that pass a test thanks to the use of immunizing devices. Now suppose a scientist (Galileo, for example) comes across a fact (say the eccentricity of the orbit of Venus) which contradicts his theory (say the Copernican system); suppose also that he admits such a contradiction, but that he takes it as a minor anomaly. The question is: should he accept the theory and adjust it in an *ad hoc* way, for this is in agreement with R_r? or should he reject the theory, for such an adjustment would violate R_a?

This antinomy might perhaps be solved by rendering the two rules more precise, for example by adding to R_a a list of devices whose use would not be compatible with a severe test; however, it is hard to believe that conflicts like these can always be evaded and it is even more difficult to believe that such a list may ever be complete.

Here a third characteristic of a methodological code crops up. Even if it were precise and self-consistent, a methodological code cannot enumerate all the circumstances in which it may be used or its validity suspended. Methodological rules are like juridical norms: they implicitly contain what may be called an "*unless*-clause". "Do not introduce *ad hoc* hypotheses unless *x*"; "do not accept weakly confirmed theories unless *y*"; "reject falsified theories, unless *z*", and so on, where the variables *x*,*y*,*z* stand for unexpected and unpredictable circumstances which it is at the discretion of the working scientist to evaluate.[8]

It might be argued it is precisely the aim of methodology to replace these variables with constants which indicate specific and well-defined circumstances. Sophisticationists think so; but two facts should warn us that this aim is neither feasible nor desirable. Not feasible, because relevant margins of ambiguity and imprecision always remain even in the most accurate code; as regards legal codes, Hobbes had already remarked that all laws are ambiguous, so short laws may be easily misunderstood while long laws are even more obscure. And not desirable, because the old saying *summum jus, summa iniuria* holds good in science, too.[9] Feyerabend had great fun in playing a *jeu de massacre* with the best methodological rules, convincingly showing that in many important cases a literal, not a creative, application of them would have eliminated the best theories.

The most relevant negative analogy between the judge and the scientist sheds further light on the limitations of methodology. A judge is only called upon to *apply* a given code; his margins of discretion and freedom, however relevant, remain confined within the limits of application. He can neither suspend the validity of the norms in force nor invent new norms; he cannot even refuse to judge if he finds the code is incomplete. But a scientist is in a different position; not only does he apply the rules of the received code, he may also *change* some of these rules and, in exceptional cases which it is in his power alone to assess, he may even question the entire code, or at least the central prescriptions of the code. Thus, while a judge is an official, the scientist is both a legislator and executor: he carries out research while, at the same time, inventing the rules of inquiry. Compared with the judge, his freedom is enormously greater.

Are there ways of reducing this freedom? My answer is that there are, but relevant margins of discretion will always escape the most sophisticated police of any Big Brother of methodology. This is a fact; and since it is a fact, we should not try to evade it. On the contrary, we should accept it and make every effort to transform it into a positive, illuminating event. The question then arises as to how science can be a rational enterprise even if it turns out to be impossible to consider it as "knowledge without a knowing subject", submitted to impersonal, universal rules.

3. FROM METHODOLOGICAL RULES TO RHETORICAL TECHNIQUES

Before entering into this question, I wish to make it clear what exactly the aforegoing considerations intend to prove.

My view is not that there are no rules or no effective rules in science. Nor do I maintain a scientist can be such an opportunist as to violate all the rules at his convenience or so free as to invent new rules any time he likes. On the contrary, a scientist is bound by rules in the same way a judge is bound by laws. These rules are imposed on the scientist by the kind of enterprise he is involved in and they can be primarily justified on the basis of the cognitive goals he wants to achieve. Moreover, in most cases, the rules are operative and give rise to univocal decisions.

What I intend to prove, on the contrary, is that, in science, there are no rules which are so precise as to eliminate personal factors or make them redundant. As regards the judge, Perelman has put it very clearly:

> "It can hardly be the case that the sole consideration of the text is enough to determine his decision: very often, the judge ought to rely his argument on the spirit of the

> system, that is on the set of values the legislator has tried to protect and promote".[10]

And, in a well-known passage, Kuhn has voiced the same opinion concerning the scientist:

> "What I am denying is neither the existence of good reasons nor that good reasons are of the sort usually described. I am, however, insisting that such reasons constitute values to be used in making choices rather than rules of choice. Scientists who share them may nevertheless make different choices in the same concrete situation".[11]

As a result of this state of affairs, which stems directly from the intrinsic limitations of methodology stated above, the rationality of science can no longer be conceived as depending on "conforming to rules". Suppose a scientist works within a research tradition and decides to stick to its methodological code. Since he cannot be completely determined by this code, for he must interpret its rules, complete them and solve their possible antinomies, the question whether he is rational or not depends on the kind of reasons he offers for the decisions he takes. Suppose now a scientist decides to abandon the current methodological code, or some relevant parts of it, and to espouse a new code. In this case, it is even more clear that his rationality cannot be evaluated in the light of either code but on the basis of the reasons he puts forward for his decision. If the question of rationality depended on conformity to a code, a theoretical change would be rational or irrational by definition, according to whether such a change is considered from the old or the new point of view.

But, it might be argued, giving reasons has to do with logic and giving good reasons has to do with standards or rules of logic. Since the limitations of methodology prove there can be no universal and impersonal standards, are we not back to square one? My reply is no. But in order to support it, I must show that we should evade that imperialism of logic and methodology to which we have been accustomed by the Cartesian tradition and thanks to which we uncritically believe that outside that empire there is no hope for rationality.

This is a difficult task, indeed. But perhaps it may be rendered less arduous if we start by noting that in science the scope for giving reasons, that is the scope for scientific reasoning, is much wider than has been thought so far by the dreamers of methodology or "logic of discovery". What they have been willing to admit into methodology are only deductive and inductive arguments obeying formal rules or universal standards; but it is clear that we need something more specific and circumstantial if we want to cover the entire field of scientific reasoning and give an account and an evaluation of all the epistemic decisions a scientist, in the multiform concrete situations of his research, is called upon to take.

Here, too, the analogy with law turns out to be illuminating. When judges apply the law or lawyers plead a case, they do not merely rely on deductive logic (practical syllogism) or inductive logic (analogy, for example); they also resort to a large stock of argumentative forms which make up what is usually called "juridical rhetoric" or "juridical dialectics". Such arguments as *a pari*, *a contrario*, *a fortiori*, *a majori ad minus*, etc., that is arguments which are persuasive but not compelling according to the standards of the other two logics, are indispensable tools for judges and lawyers. Even more so for scientists.

Let us consider a typical situation. Suppose that certain observational consequences O may be derived from a theory T and that O really happen. Then, a scientist makes use of deduction when he infers O from T; such an inference is disciplined by the rules of formal logic and mathematics. Moreover, the scientist makes use of induction when, for example, he affirms that, given O, T is probable or when he claims that, since O is unexpected without T, T is more probable. These inferences are disciplined (if they are at all) by the canons of inductive logic and are sometimes formalized by theorems of probability calculus. Lastly, the scientist resorts to arguments of a different kind when, for example, he argues that T has to be accepted because, say, it is more unifying, or T is worth working with because, say, it is more promising; or O is relevant, severe, novel, etc. because it possesses such and such properties; and so on. Arguments to such ends are neither deductive nor inductive for they are neither formally conclusive nor empirically stringent. Rather, they are dialectical or rhetorical; they aim at persuading and convincing by offering good reasons in situations in which there can be no better proof.

Ever since Descartes, methodology has been conceived with the aim of plugging up the holes of formal logic and covering the whole domain of argumentation, on the assumption that, if that were not possible, then science would loose its rationality. But that is *not* possible and, nonetheless, science *is* rational. Therefore, the best thing to do is to reject such an assumption and get rid of the Cartesian syndrome. If we aim at a realistic image of science, we should stop sophisticating the methodological rules and improve our understanding of scientific rhetoric instead. As for methodology, we should do what Perelman and his school have been doing with formal logic and the Cartesian tradition.[12] As I see it, the programme would have to be articulated into different phases which I can only roughly outline here.

We should start by reviewing the real arguments scientists make use of, without worrying if they fit into the categories of our favourite logic or methodology. Then we should specify the patterns of these arguments and classify them; we should also examine the modes of their use and the contexts in which they are used, fix their goals and establish the factors on which their efficacy depends. This is quite a different job to sophisticating the rules: the one leads to an impersonal, a-historical, content-independent methodology which is always at odds with the great variety of argumentative forms actually used by scientists and which will never be able to discipline them or give a rationale for such discipline; the other may lead to a substantive, situational logic, comprehensive of such forms, close to the scientist's work.

Ever since the breakdown of juridical positivism, philosophers of law have been increasingly interested in studying the procedures actually used by legislators, judges and lawyers and have extended their set of tools to include the theory of argumentation and other resources, besides formal logic. The time is ripe, in my view, to do the same in the philosophy of science.

In the last part of this paper, I wish to give a rough sketch of a case of scientific rhetoric and draw some conclusions as a tentative moral.

4. SCIENTIFIC ARGUMENTATION AT WORK. DARWIN'S RHETORIC

At the beginning of the last chapter of the *Origin*, Darwin wrote that "this volume is one long argument".[13] The validity of Darwin's chain of arguments and its ability to do the job for which he had introduced it, have been disputed many times by his critics ever since the book first came out. Darwin admitted that some of his arguments were "vague" and "conjectural"; but, on the whole, he was confident that they were convincing, at least for those who were willing to rise above the "blindness of preconceived opinion".[14] He also proved to be "fully convinced" of the "truth" of his views;[15] "it can hardly be supposed", he wrote, "that a false theory would explain, in so satisfactory a manner as does the theory of natural selection, the several large classes of facts above specified".[16]

In passages like these, Darwin expresses his confidence in the rules of the hypothetico-deductive method (in particular, in the "consilience of inductions") and accordingly he proves to be of the opinion that facts and theory are all we need to play the game of science. But the practice of his research was quite different from his methodological creed. The point is precisely that the hypothetico-deductive method itself was disputed and that Darwin's theory faced many problems. To solve these problems, arguments relying solely on the relationship between the theory and the facts - that is, deductive or inductive arguments - could not be sufficient. And thus, quite typically, Darwin resorted to a large stock of rhetorical arguments. His book is full of them and examination of its rhetorical apparatus would provide ample information on how scientific research is actually pursued. Let us choose a small, random sample.

One problem for Darwin's theory was that the idea of "natural selection" seemed to be metaphysical or speculative and thus without any causal efficiency. To this he replied:

> [1] In the literal sense of the word, no doubt, natural selection is a false term; but who ever objected to chemists speaking of the elective affinities of the various elements - and yet an acid cannot strictly be said to elect the base with which it in preference combines. It has been said that I speak of natural selection as an active power of Deity; but who objects to an author speaking of the attraction of gravity as ruling the movements of the planetes?[17]

This is both an *argument of reciprocity*, with which the same treatment is invoked for situations asserted to be similar, and an *argument of retortion*, with which an objection is rejected on the grounds that, if it were accepted, a considerable amount of admittedly good scientific practice would have to be rejected, too. The dialectical strength of this argument obviously depends on the extent to which the authority of the practice in question is recognized by the scientific community and on how the alleged similarity of situations is admitted.

Another problem for Darwin's theory was that the hypothetico-deductive method proves only indirectly and that, in any case, he simply derived the fact of the variation of species without explaining its essence. Darwin's reply to this is another argument of retortion:

> [2] It has recently been objected that this is an unsafe method of arguing; but it is a method used in judging of the common events of life, and has often been used by the greatest natural philosophers. The undulatory theory of light has thus been arrived at; and the belief in the revolution of

> the earth on its own axis was until lately supported by hardly any direct evidence. It is no valid objection that science as yet throws no light on the far higher problem of the essence or origin of life. Who can explain what is the essence of the attraction of gravity? No one now objects to following out the results consequent on this unknown element of attraction; notwithstanding that Leibniz formerly accused Newton of introducing "occult qualities and miracles into philosophy".[18]

A third objection to Darwin's theory is that natural selection is implausible as an efficient cause of all the observed phenomena of variation. To this he replied with the two following arguments:

> [3] As man can produce, and certainly has produced, a great result by his methodical and unconscious means of selection, what may not natural selection effect? Man can act only on external and visible characters: Nature, if I may be allowed to personify the natural preservation or survival of the fittest, cares nothing for appearences, except in so far as they are useful to any being. She can act on every internal organ, on every shade of constitutional difference, on the whole machinery of life. Man selects only for his own good: Nature only for that of the being which she tends. ... Can we wonder, then, that Nature's productions should be far "truer" in character than man's productions; that they should be infinitely better adapted to the most complex conditions of life, and should plainly bear the stamp of far higher workmanship?[19]

> [4] If man can by patience select variations useful to him, why, under changing and complex conditions of life, should not variations useful to nature's living products often arise, and be preserved or selected? What limit can be put to this power, acting during long ages and rigidly scrutinising the whole constitution, structure, habits of each creature, - favouring the good and rejecting the bad?[20]

Both these arguments present the schema of the *argument of double hierarchy*: if *A* (= breeder or craftsman) can produce *a* (= domestic variations or artifacts), why could not *B* (= Nature), which is so superior to *A*, produce *b* (= formation of species or complex organs)? The argument is convincing, although it raises just that problem which Darwin tries to evade, namely the personification of Nature with all the disputable consequences, philosophical, theological, etc., such a personification leads to.

A more serious challenge to Darwin's theory came from its admitted empirical anomalies. How can it explain such complex organs as the eye? And how can it give an account of those organs which are apparently harmful to their bearers? To the first question Darwin replied:

> [5] He who will go thus far, ought not to hesitate to go one step farther, if he finds on finishing this volume that large bodies of facts, otherwise inexplicable, can be explained by the theory of modification through natural selection; he ought to admit that a structure even as perfect as an eagle's eye might thus be formed, although in this case he does not know the transitional states.[21]

Darwin's answer to the second question is the following:

> [6] It is admitted that the rattlesnake has a poisonfang for its own defence, and for the destruction of its prey; but some authors suppose that at the same time it is furnished with a rattle for its own injury, namely to warn its prey. I would almost as soon believe that the cat curls the end of its tail when preparing to spring, in order to warn the doomed mouse.[22]

The first reply is an *argument of pressing on*. It tries to convince the interlocutor that the theory of natural selection has to be pursued in order to get future results and not to invalidate those already acquired. But clearly it is more an article of faith than an appeal to reason, for Darwin himself admits that he lacks empirical evidence (the "transitional states"). *Allez en avant, l'entendence suivra*. The second reply is an *argument from ridicule*: supposing it is really an argument, it does not confute the objection as such, rather it aims at discrediting it. As a refutation of an objection, the argument is as effective as a witticism; it too amounts to an invitation to carry on with the theory in the hope that future results will support it better.

Another very serious problem for Darwin is that his theory also faced external conceptual problems. As is well known, a grave objection to natural selection came from physics and concerned the great lapse of time the theory had to assume in order to explain the evolution of species. To this objection, Darwin replied:

> [7] With respect to the lapse of time not having been sufficient since our planet was consolidated for the assumed

> amount of organic change, and this objection, as urged by Sir William Thompson, is probably one of the gravest as yet advanced, I can only say, firstly, that we do not know at what rate species change as measured by years, and secondly, that many philosophers are not as yet willing to admit that we know enough of the constitution of the universe and of the interior of our globe to speculate with safety on its past duration.[23]

This is an *argument ad ignorantiam*. Darwin does not offer any positive empirical evidence to rebut Thompson's objection; nor does he maintain that Thompson's objection is not stringent. However, he tends to transform an *objection against* into an *argument for*; he relies on the ignorance surrounding his opponent's view, arguing that, since this view has not yet been conclusively established, then his own theory has to be accepted. Here, too, he construes a typically rhetorical alternative: either his opponent proves that the theory is false or it has to be accepted as true.

Other problems with Darwin's theory came from the fact that alternative hypotheses could be invoked to explain the same phenomena. Here is one of the ways in which he rejected the idea of sudden transformations:

> [8] He who believes that some ancient form was transformed suddenly through an internal force or tendency into, for instance, one furnished with wings, will be almost compelled to assume, in opposition to all analogy, that many individuals varied *simultaneously*. It cannot be denied that such abrupt and great changes of structure are widely different from those which most species apparently have undergone. He will further be compelled to believe that many structures beautifully adapted to all the other parts of the same creature and to the surrounding conditions, have been suddenly produced; and of such complex and wonderful co-adaptations, he will not be able to assign a shadow of an explanation. He will be forced to admit that these great and sudden transformations have left no trace of their action on the embryo. To admit all this is, as it seems to me, to enter into the realms of miracle, and to leave those of science.[24]

This is an *argument ad hominem*. Darwin does not confute the thesis itself with positive, empirical evidence; rather, he makes use of propositions he believes his opponent should accept. The move consists in considering these propositions as unintentional and undesirable consequences of the opponent's thesis, and in confronting them with other views he shares or must share if he wants to pursue scientific research. The alternative Darwin intends to construe is then clear: either the opponent must revise common principles of science or accepted results or he must discard his own thesis.

To the idea of independent creation of species he reacted in this way:

> [9] He who believes that each equine species was independently created, will, I presume, assert that each species has been created with a tendency to vary, both under nature and under domestication, in this particular manner, so as often to become striped like the other species of the genus; and that each has been inhabiting distant quarters of the world, to produce hybrids resembling in their stripes, not their own parents, but other species of the genus. To admit this view is, as it seems to me, to reject a real for an unreal, or at least for an unknown, cause. It makes the works of God a mere mockery and deception; I would almost as soon believe the old and ignorant cosmologists, that fossil shells had never lived, but had been created in stone so as to mock the shells living on the sea-shore.[25]

Here too, Darwin rebuts a possible, rival explanation not by giving positive evidence against it but by trying to put his opponent in the situation of either contradicting certain theses he is supposed to admit or of accepting Darwin's own explanation. In so doing, Darwin typically forces the situation: his opponent's alleged cause is first declared "unreal", then "unknown" and finally ridiculed as a "mockery".

What conclusions can we draw from this small sample of dialectical arguments? The following theses may perhaps be considered as part of a tentative moral.

First, Darwin expresses his adherence to a method of inquiry (the hypothetico-deductive method) and does stick to some of its rules. For example, he considers his theory confirmed for it explains a great variety of facts (rule of the consilience of inductions); or he rejects alternative hypotheses on the ground that they have some contrary evidence (rule of rejection through falsification of the consequences). But the rules of Darwin's method have an open texture, and in any case Darwin takes them as open to different interpretations. For this reason, he does not rebut the theory of natural selection just because it is faced with empirical difficulties or it is contradicted by other accepted theories; rather, he makes use of his margins of discretion to interpret the content of the rules of his method in a loose, tolerant, and, to him, more convenient way.

Thus, second, to accredit these interpretations, Darwin must resort to rhetorical arguments. Just like a judge or, better, like a lawyer, he has to argue that certain interpretations are admissible, fruitful and suitable for the case in question, while others are not. Since he cannot prove his theory in an impersonal way by using facts alone, for the theory is underdetermined by the facts or even undermined by them, he can only make use of persuasive arguments aimed at inducing agreement on his line of research.

Third, Darwin's rhetoric performs different functions; with reference to our small sample of arguments, these are: to vindicate the hypothetico-deductive method as a legitimate procedure ([1] and [2]); to suggest that the theory of natural selection is intrinsically plausible and worth attention ([3] and [4]); to invite those who find it so to carry on with it in spite of its difficulties ([5]); to minimize the importance of these difficulties ([6]); to refute or discredit rival theories ([7] and [8]); and to reject objections to the theory ([9]).

Fourth, Darwin's rationality does not depend on his method, because the method itself is disputed as a way of proving and because, even taking it for granted, its rules do not fully discipline his decisions. Rather, Darwin's rationality depends on how strong his arguments are and on how good and efficient his reasons are for the goals they are devised to achieve. There are no universal or impersonal rules to evaluate this strength and this goodness; all depends on the criticism which, in a given situation, is put forward against the arguments and on how the arguments stand up to this criticism. We may say that that decision or behaviour is rational which is supported by strong arguments or good reasons and that argument is strong or that reason is good which best faces all kinds of criticism.

One final remark. It might be objected that rhetorical arguments are, at the very best, temporary moves which one can make use of in the middle of a dispute, but which will be superseded when new evidence is acquired. "Ultimately", as Popper says in the passage I quoted at the beginning, the fate of a theory is decided by the result of a test. I agree with the spirit of this golden rule, but I should like to point out that a test is a test only if we decide it is. Does not Popper, and especially Lakatos, acknowledge that it is always possible to appeal against a fact? Moreover, what does "ultimately" mean? In science, we are never in a final situation and there is no Last Judgement. This is not to say that facts play no role and that, at a certain point, further appeal against them does not become unreasonable. But to establish when that point has been reached is a question which can be decided upon case by case, not with reference to a table of Universal Commandments.

Summing up: Moses is dead, but fortunately there are other ways of crossing the Desert. The fact that these are more tentative and less safe does not make them less promising or less attractive.

Department of Philosophy
University of Pisa

NOTES

I would like to express my gratitude to the participants at the Conference for their comments. I owe a special debt of gratitude to Aristides Baltas for reading the first draft of this paper and suggesting how to make it better than it is in its present version.

1. K. Popper, *The Logic of Scientific Discovery*, Hutchinson, London 1959, p.109.
2. See K. Popper, *Objective Knowledge. An Evolutionary Approach*, Clarendon Press Oxford, ch.3.
3. G. Galilei, *Dialogue Concerning the two Chief World Systems – Ptolemaic and Copernican*, trans. S. Drake, Berkeley and Los Angeles 1967, p. 53 (marginal annotation).
4. This is Feyerabend's case. See, for example, his *Against Method*, New Left Books, London 1975, p. 154.
5. See my papers "Narcissus at the Pool: Scientific Method and the History of Science", forthcoming in *Organon* 1986; "Methodological Sophisticationism: A Degenerating Programme", read at the Lakatos Memorial Conference on "Criticism and the Growth of Knowledge: 20 Years After", Thessaloniki, 18–22 August 1986, to be published in the Proceedings.
6. See I. Lakatos, "Falsification and the Methodology of Scientific Research Programmes", in *Philosophical Papers*, 2 vols, edited by J. Worrall and G. Currie, Cambridge University Press, Cambridge 1978, vol. 1, p. 31 and p. 91; "History of Science and its Rational Reconstructions", in *Op. cit.*, p. 103.
7. See my paper "From Methodology to Argumentation. A Post-Cartesian Approach to Scientific Rationality", read at the Four Societies Meeting, Pittsburgh 23-26 October 1986, in *PSA* 1986, volume 2.
8. With reference to juridical codes, Perelman says that "no

text, however precise, can enumerate all the situations, often unforeseen, in which it will cease to be applicable. At the very best, it will contain such clauses as 'cause beyond control','invincible force of events','extraordinary situation' which limit its application, but, in the last resort, it is the judge or the policeman who must reinterpret them in each concrete situation". See Ch. Perelman, "Droit, philosophie et argumentation" in *Handelingen van de Vereniging voor Wijsbegeerte des Rechtes*, LV, 1971, pp. 3-21.

9. See Ch. Perelman, "Droit, logique et argumentation", in *Le Champ de l'argumentation*, P.U.B., Bruxelles 1970, p. 143: "in law, the use of vague notions is not necessarily a defect". It is interesting to note that, immediately after, Perelman adds that this does not hold in science.
10. Ch. Perelman, "Droit, philosophie et argumentation", *cit.*
11. T. Kuhn, "Reflections of My Critics", in I. Lakatos, A. Musgrave (eds.), *Criticism and the Growth of Knowledge*, Cambridge University Press, Cambridge 1970, p. 262.
12. Perelman, however, is not willing to admit much rhetoric into scientific contexts. See, for instance, his critical examination of Polanyi's *Personal Knowledge*, in *Le champ de l'argumentation*, *cit.* ch. XXVI. Sometimes Perelman seems to confine rhetorical arguments to the "context of discovery"; he says, for instance, that "in experimental science, their role concerns more the invention of hypotheses than the proof of their foundation" (*ibidem*, p. 117). On other occasions, argumentation is said to concern such questions as the introduction of a "not satisfactory theory" (*ibidem*, p. 349) or the decision to integrate new facts into the field of research. See *The Idea of Justice and the Problem of Argument*, trans. J. Petrie, Routledge, London 1963, p. 92.
13. C. Darwin, *The Origin of Species*, Everyman's Library, London 1982, p. 437.
14. *Op. cit.*, p. 457.
15. Op. cit., p. 456.
16. Op. cit., p. 455.
17. Op. cit., p. 81.
18. Op. cit., p. 455.
19. Op. cit., p. 83.
20. Op. cit., p. 445.
21. Op. cit., p. 168.
22. Op. cit., pp. 186-87.
23. Op. cit., pp. 442-43.
24. Op. cit., p. 229.
25. Op. cit., pp. 153-54.

INDEX OF NAMES

SYNTHESE LIBRARY

Studies in Epistemology, Logic, Methodology,
and Philosophy of Science

1. J. M. Bochénski, *A Precis of Mathematical Logic.* 1959.
2. P. L. Guiraud, *Problèmes et méthodes de la statistique linguistique.* 1960.
3. Hans Freudenthal (ed.), *The Concept and the Role of the Model in Mathematics and Natural and Social Sciences.* 1961.
4. Evert W. Beth, *Formal Methods. An Introduction to Symbolic Logic and the Study of Effective Operations in Arithmetic and Logic.* 1962.
5. B. H. Kazemier and D. Vuysje (eds.), *Logic and Language. Studies Dedicated to Professor Rudolf Carnap on the Occasion of His Seventieth Birthday.* 1962.
6. Marx W. Wartofsky (ed.), *Proceedings of the Boston Colloquium for the Philosophy of Science 1961–1962.* Boston Studies in the Philosophy of Science, Volume I. 1963.
7. A. A. Zinov'ev, *Philosophical Problems of Many-Valued Logic.* 1963.
8. Georges Gurvitch, *The Spectrum of Social Time.* 1964.
9. Paul Lorenzen, *Formal Logic.* 1965.
10. Robert S. Cohen and Marx W. Wartofsky (eds.), *In Honor of Philipp Frank.* Boston Studies in the Philosophy of Science, Volume II. 1965.
11. Evert W. Beth, *Mathematical Thought. An Introduction to the Philosophy of Mathematics.* 1965.
12. Evert W. Beth and Jean Piaget, *Mathematical Epistemology and Psychology.* 1966.
13. Guido Küng, *Ontology and the Logistic Analysis of Language. An Enquiry into the Contemporary Views on Universals.* 1967.
14. Robert S. Cohen and Marx W. Wartofsky (eds.), *Proceedings of the Boston Colloquium for the Philosophy of Sciences 1964–1966. In Memory of Norwood Russell Hanson.* Boston Studies in the Philosophy of Science, Volume III. 1967.
15. C. D. Broad, *Induction, Probability, and Causation. Selected Papers.* 1968.
16. Günther Patzig, *Aristotle's Theory of the Syllogism. A Logical-Philosophical Study of Book A of the Prior Analytics.* 1968.
17. Nicholas Rescher, *Topics in Philosophical Logic.* 1968.
18. Robert S. Cohen and Marx W. Wartofsky (eds.), *Proceedings of the Boston Colloquium for the Philosophy of Science 1966–1968.* Boston Studies in the Philosophy of Science, Volume IV. 1969
19. Robert S. Cohen and Marx W. Wartofsky (eds.), *Proceedings of the Boston Colloquium for the Philosophy of Science 1966–1968.* Boston Studies in the Philosophy of Science, Volume V. 1969

20. J. W. Davis, D. J. Hockney, and W. K. Wilson (eds.), *Philosophical Logic.* 1969
21. D. Davidson and J. Hintikka (eds.), *Words and Objections. Essays on the Work of W. V. Quine.* 1969.
22. Patrick Suppes. *Studies in the Methodology and Foundations of Science. Selected Papers from 1911 to 1969.* 1969
23. Jaakko Hintikka, *Models for Modalities. Selected Essays.* 1969
24. Nicholas Rescher *et al.* (eds.), *Essays in Honor of Carl G. Hempel. A Tribute on the Occasion of His Sixty-Fifth Birthday.* 1969
25. P. V. Tavanec (ed.), *Problems of the Logic of Scientific Knowledge.* 1969
26. Marshall Swain (ed.), *Induction, Acceptance, and Rational Belief.* 1970.
27. Robert S. Cohen and Raymond J. Seeger (eds.), *Ernst Mach: Physicist and Philosopher.* Boston Studies in the Philosophy of Science, Volume VI. 1970.
28. Jaakko Hintikka and Patrick Suppes, *Information and Inference.* 1970.
29. Karel Lambert, *Philosophical Problems in Logic. Some Recent Developments.* 1970.
30. Rolf A. Eberle, *Nominalistic Systems.* 1970.
31. Paul Weingartner and Gerhard Zecha (eds.), *Induction, Physics, and Ethics.* 1970.
32. Evert W. Beth, *Aspects of Modern Logic.* 1970.
33. Risto Hilpinen (ed.), *Deontic Logic: Introductory and Systematic Readings.* 1971.
34. Jean-Louis Krivine, *Introduction to Axiomatic Set Theory.* 1971.
35. Joseph D. Sneed, *The Logical Structure of Mathematical Physics.* 1971.
36. Carl R. Kordig, *The Justification of Scientific Change.* 1971.
37. Milic Capek, *Bergson and Modern Physics.* Boston Studies in the Philosophy of Science, Volume VII. 1971.
38. Norwood Russell Hanson, *What I Do Not Believe, and Other Essays* (ed. by Stephen Toulmin and Harry Woolf). 1971.
39. Roger C. Buck and Robert S. Cohen (eds.), *PSA 1970. In Memory of Rudolf Carnap.* Boston Studies in the Philosophy of Science, Volume VIII. 1971
40. Donald Davidson and Gilbert Harman (eds.), *Semantics of Natural Language.* 1972.
41. Yehoshua Bar-Hillel (ed.), *Pragmatics of Natural Languages.* 1971.
42. Sören Stenlund, *Combinators, λ-Terms and Proof Theory.* 1972.
43. Martin Strauss, *Modern Physics and Its Philosophy. Selected Paper in the Logic, History, and Philosophy of Science.* 1972.
44. Mario Bunge, *Method, Model and Matter.* 1973.
45. Mario Bunge, *Philosophy of Physics.* 1973.
46. A. A. Zinov'ev, *Foundations of the Logical Theory of Scientific Knowledge (Complex Logic).* (Revised and enlarged English edition with an appendix by G. A. Smirnov, E. A. Sidorenka, A. M. Fedina, and L. A. Bobrova.) Boston Studies in the Philosophy of Science, Volume IX. 1973.
47. Ladislav Tondl, *Scientific Procedures.* Boston Studies in the Philosophy of Science, Volume X. 1973.
48. Norwood Russell Hanson, *Constellations and Conjectures* (ed. by Willard C. Humphreys, Jr.). 1973
49. K. J. J. Hintikka, J. M. E. Moravcsik, and P. Suppes (eds.), *Approaches to Natural Language.* 1973.
50. Mario Bunge (ed.), *Exact Philosophy – Problems, Tools, and Goals.* 1973.
51. Radu J. Bogdan and Ilkka Niiniluoto (eds.), *Logic, Language, and Probability.* 1973.
52. Glenn Pearce and Patrick Maynard (eds.), *Conceptual Change.* 1973.

53. Ilkka Niiniluoto and Raimo Tuomela, *Theoretical Concepts and Hypothetico-Inductive Inference.* 1973.
54. Roland Fraissé, *Course of Mathematical Logic* – Volume 1: *Relation and Logical Formula.* 1973.
55. Adolf Grünbaum, *Philosophical Problems of Space and Time.* (Second, enlarged edition.) Boston Studies in the Philosophy of Science, Volume XII. 1973.
56. Patrick Suppes (ed.), *Space, Time, and Geometry.* 1973.
57. Hans Kelsen, *Essays in Legal and Moral Philosophy* (selected and introduced by Ota Weinberger). 1973.
58. R. J. Seeger and Robert S. Cohen (eds.), *Philosophical Foundations of Science.* Boston Studies in the Philosophy of Science, Volume XI. 1974.
59. Robert S. Cohen and Marx W. Wartofsky (eds.), *Logical and Epistemological Studies in Contemporary Physics.* Boston Studies in the Philosophy of Science, Volume XIII. 1973.
60. Robert S. Cohen and Marx W. Wartofsky (eds.), *Methodological and Historical Essays in the Natural and Social Sciences. Proceedings of the Boston Colloquium for the Philosophy of Science 1969–1972.* Boston Studies in the Philosophy of Science, Volume XIV. 1974.
61. Robert S. Cohen, J. J. Stachel, and Marx W. Wartofsky (eds.), *For Dirk Struik. Scientific, Historical and Political Essays in Honor of Dirk J. Struik.* Boston Studies in the Philosophy of Science, Volume XV. 1974.
62. Kazimierz Ajdukiewicz, *Pragmatic Logic* (transl. from the Polish by Olgierd Wojtasiewicz). 1974.
63. Sören Stenlund (ed.), *Logical Theory and Semantic Analysis. Essays Dedicated to Stig Kanger on His Fiftieth Birthday.* 1974.
64. Kenneth F. Schaffner and Robert S. Cohen (eds.), *Proceedings of the 1972 Biennial Meeting, Philosophy of Science Association.* Boston Studies in the Philosophy of Science, Volume XX. 1974.
65. Henry E. Kyburg, Jr., *The Logical Foundations of Statistical Inference.* 1974.
66. Marjorie Grene, *The Understanding of Nature. Essays in the Philosophy of Biology.* Boston Studies in the Philosophy of Science, Volume XXIII. 1974.
67. Jan M. Broekman, *Structuralism: Moscow, Prague, Paris.* 1974.
68. Norman Geschwind, *Selected Papers on Language and the Brain,* Boston Studies in the Philosophy of Science, Volume XVI. 1974.
69. Roland Fraissé, *Course of Mathematical Logic* – Volume 2: *Model Theory.* 1974.
70. Andrzej Grzegorczyk, *An Outline of Mathematical Logic. Fundamental Results and Notions Explained with All Details.* 1974.
71. Franz von Kutschera, *Philosophy of Language.* 1975.
72. Juha Manninen and Raimo Tuomela (eds.), *Essays on Explanation and Understanding. Studies in the Foundations of Humanities and Social Sciences.* 1976.
73. Jaakko Hintikka (ed.), *Rudolf Carnap, Logical Empiricist. Materials and Perspectives.* 1975.
74. Milic Capek (ed.), *The Concepts of Space and Time. Their Structure and Their Development.* Boston Studies in the Philosophy of Science, Volume XXII. 1976.
75. Jaakko Hintikka and Unto Remes, *The Method of Analysis. Its Geometrical Origin and Its General Significance.* Boston Studies in the Philosophy of Science, Volume XXV. 1974.
76. John Emery Murdoch and Edith Dudley Sylla, *The Cultural Context of Medieval Learning.* Boston Studies in the Philosophy of Science, Volume XXVI. 1975.

77. Stefan Amsterdamski, *Between Experience and Metaphysics. Philosophical Problems of the Evolution of Science.* Boston Studies in the Philosophy of Science, Volume XXXV. 1975.
78. Patrick Suppes (ed.), *Logic and Probability in Quantum Mechanics.* 1976.
79. Hermann von Helmholtz: *Epistemological Writings. The Paul Hertz/Moritz Schlick Centenary Edition of 1921 with Notes and Commentary by the Editors.* (Newly translated by Malcolm F. Lowe. Edited, with an Introduction and Bibliography, by Robert S. Cohen and Yehuda Elkana.) Boston Studies in the Philosophy of Science, Volume XXXVII. 1975.
80. Joseph Agassi, *Science in Flux.* Boston Studies in the Philosophy of Science, Volume XXVIII. 1975.
81. Sandra G. Harding (ed.), *Can Theories Be Refuted? Essays on the Duhem-Quine Thesis.* 1976.
82. Stefan Nowak, *Methodology of Sociological Research. General Problems.* 1977.
83. Jean Piaget, Jean-Blaise Grize, Alina Szeminska, and Vinh Bang, *Epistemology and Psychology of Functions.* 1977.
84. Marjorie Grene and Everett Mendelsohn (eds.), *Topics in the Philosophy of Biology.* Boston Studies in the Philosophy of Science, Volume XXVII. 1976.
85. E. Fischbein, *The Intuitive Sources of Probabilistic Thinking in Children.* 1975.
86. Ernest W. Adams, *The Logic of Conditionals. An Application of Probability to Deductive Logic.* 1975.
87. Marian Przelecki and Ryszard Wójcicki (eds.), *Twenty-Five Years of Logical Methodology in Poland.* 1976.
88. J. Topolski, *The Methodology of History.* 1976.
89. A. Kasher (ed.), *Language in Focus: Foundations, Methods and Systems. Essays Dedicated to Yehoshua Bar-Hillel.* Boston Studies in the Philosophy of Science, Volume XLIII. 1976.
90. Jaakko Hintikka, *The Intentions of Intentionality and Other New Models for Modalities.* 1975.
91. Wolfgang Stegmüller, *Collected Papers on Epistemology, Philosophy of Science and History of Philosophy.* 2 Volumes. 1977.
92. Dov M. Gabbay, *Investigations in Modal and Tense Logics with Applications to Problems in Philosophy and Linguistics.* 1976.
93. Radu J. Bodgan, *Local Induction.* 1976.
94. Stefan Nowak, *Understanding and Prediction. Essays in the Methodology of Social and Behavioral Theories.* 1976.
95. Peter Mittelstaedt, *Philosophical Problems of Modern Physics.* Boston Studies in the Philosophy of Science, Volume XVIII. 1976.
96. Gerald Holton and William Blanpied (eds.), *Science and Its Public: The Changing Relationship.* Boston Studies in the Philosophy of Science, Volume XXXIII. 1976.
97. Myles Brand and Douglas Walton (eds.), *Action Theory.* 1976.
98. Paul Gochet, *Outline of a Nominalist Theory of Proposition. An Essay in the Theory of Meaning.* 1980.
99. R. S. Cohen, P. K. Feyerabend, and M. W. Wartofsky (eds.), *Essays in Memory of Imre Lakatos.* Boston Studies in the Philosophy of Science, Volume XXXIX. 1976.
100. R. S. Cohen and J. J. Stachel (eds.), *Selected Papers of Léon Rosenfield.* Boston Studies in the Philosophy of Science, Volume XXI. 1978.
101. R. S. Cohen, C. A. Hooker, A. C. Michalos, and J. W. van Evra (eds.), *PSA 1974: Proceedings of the 1974 Biennial Meeting of the Philosophy of Science Association.* Boston Studies in the Philosophy of Science, Volume XXXII. 1976.

102. Yehuda Fried and Joseph Agassi, *Paranoia: A Study in Diagnosis*. Boston Studies in the Philosophy of Science, Volume L. 1976.
103. Marian Przelecki, Klemens Szaniawski, and Ryszard Wójcicki (eds.), *Formal Methods in the Methodology of Empirical Sciences*. 1976.
104. John M. Vickers, *Belief and Probability*. 1976.
105. Kurt H. Wolff, *Surrender and Catch: Experience and Inquiry Today*. Boston Studies in the Philosophy of Science, Volume LI. 1976.
106. Karel Kosík, *Dialectics of the Concrete*. Boston Studies in the Philosophy of Science, Volume LII. 1976.
107. Nelson Goodman, *The Structure of Appearance* (Third edition.) Boston Studies in the Philosophy of Science, Volume LIII. 1977.
108. Jerzy Giedymin (ed.), *Kazimierz Ajdukiewicz: The Scientific World-Perspective and Other Essays, 1931-1963*. 1978.
109. Robert L. Causey, *Unity of Science*. 1977.
110. Richard E. Grandy, *Advanced Logic for Applications*. 1977.
111. Robert P. McArthur, *Tense Logic*. 1976.
112. Lars Lindahl, *Position and Change. A Study in Law and Logic*. 1977.
113. Raimo Tuomela, *Dispositions*. 1978.
114. Herbert A. Simon, *Models of Discovery and Other Topics in the Methods of Science*. Boston Studies in the Philosophy of Science, Volume LIV. 1977.
115. Roger D. Rosenkrantz, *Inference, Method and Decision*. 1977.
116. Raimo Tuomela, *Human Action and Its Explanation. A Study on the Philosophical Foundations of Psychology*. 1977.
117. Morris Lazerowitz, *The Language of Philosophy. Freud and Wittgenstein*. Boston Studies in the Philosophy of Science, Volume LV. 1977.
119. Jerzy Pelc, *Semiotics in Poland, 1894-1969*. 1978.
120. Ingmar Pörn, *Action Theory and Social Science. Some Formal Models*. 1977.
121. Joseph Margolis, *Persons and Mind. The Prospects of Nonreductive Materialism*. Boston Studies in the Philosophy of Science, Volume LVII. 1977.
122. Jaakko Hintikka, Ilkka Niiniluoto, and Esa Saarinen (eds.), *Essays on Mathematical and Philosophical Logic*. 1978.
123. Theo A. F. Kuipers, *Studies in Inductive Probability and Rational Expectation*. 1978.
124. Esa Saarinen, Risto Hilpinen, Ilkka Niiniluoto, and Merrill Provence Hintikka (eds.), *Essays in Honour of Jaakko Hintikka on the Occasion of His Fiftieth Birthday*. 1978.
125. Gerard Radnitzky and Gunnar Andersson (eds.), *Progress and Rationality in Science*. Boston Studies in the Philosophy of Science, Volume LVIII. 1978.
126. Peter Mittelstaedt, *Quantum Logic*. 1978.
127. Kenneth A. Bowen, *Model Theory for Modal Logic. Kripke Models for Modal Predicate Calculi*. 1978.
128. Howard Alexander Bursen, *Dismantling the Memory Machine. A Philosophical Investigation of Machine Theories of Memory*. 1978.
129. Marx W. Wartofsky, *Models: Representation and the Scientific Understanding*. Boston Studies in the Philosophy of Science, Volume XLVIII. 1979.
130. Don Ihde, *Technics and Praxis. A Philosophy of Technology*. Boston Studies in the Philosophy of Science, Volume XXIV. 1978.
131. Jerzy J. Wiatr (ed.), *Polish Essays in the Methodology of the Social Sciences*. Boston Studies in the Philosophy of Science, Volume XXIX. 1979.
132. Wesley C. Salmon (ed.), *Hans Reichenbach: Logical Empiricist*. 1979.

133. Peter Bieri, Rolf-P. Horstmann, and Lorenz Krüger (eds.), *Transcendental Arguments in Science Essays in Epistemology*. 1979.
134. Mihailo Marković and Gajo Petrović (eds.), *Praxis, Yugoslav Essays in the Philosophy and Methodology of the Social Sciences*. Boston Studies in the Philosophy of Science, Volume XXXVI. 1979.
135. Ryszard Wójcicki, *Topics in the Formal Methodology of Empirical Sciences*. 1979.
136. Gerard Radnitzky and Gunnar Andersson (eds.), *The Structure and Development of Science*. Boston Studies in the Philosophy of Science, Volume LIX. 1979.
137. Judson Chambers Webb, *Mechanism, Mentalism, and Metamathematics. An Essay on Finitism*. 1980.
138. D. F. Gustafson and B. L. Tapscott (eds.), *Body, Mind, and Method. Essays in Honor of Virgil C. Aldrich*. 1979.
139. Leszek Nowak, *The Structure of Idealization. Towards a Systematic Interpretation of the Marxian Idea of Science*. 1979.
140. Chaim Perelman, *The New Rhetoric and the Humanities. Essays on Rhetoric and Its Applications*. 1979.
141. Wlodzimierz Rabinowicz, *Universalizability. A Study in Morals and Metaphysics*. 1979.
142. Chaim Perelman, *Justice, Law, and Argument. Essays on Moral and Legal Reasoning*. 1980.
143. Stig Kanger and Sven Öhman (eds.), *Philosophy and Grammar. Papers on the Occasion of the Quincentennial of Uppsala University*. 1980.
144. Tadeusz Pawlowski, *Concept Formation in the Humanities and the Social Sciences*. 1980.
145. Jaakko Hintikka, David Gruender, and Evandro Agazzi (eds.), *Theory Change, Ancient Axiomatics, and Galileo's Methodology. Proceedings of the 1978 Pisa Conference on the History and Philosophy of Science*, Volume I. 1981.
146. Jaakko Hintikka, David Gruender, and Evandro Agazzi, *Probabilistic Thinking, Thermodynamics, and the Interaction of the History and Philosophy of Science. Proceedings of the 1978 Pisa Conference on the History and Philosophy of Science*, Volume II. 1981.
147. Uwe Mönnich (ed.), *Aspects of Philosophical Logic. Some Logical Forays into Central Notions of Linguistics and Philosophy*. 1981.
148. Dov M. Gabbay, *Semantical Investigations in Heyting's Intuitionistic Logic*. 1981.
149. Evandro Agazzi (ed.), *Modern Logic – A Survey. Historical, Philosophical, and Mathematical Aspects of Modern Logic and Its Applications*. 1981.
150. A. F. Parker-Rhodes, *The Theory of Indistinguishables. A Search for Explanatory Principles below the Level of Physics*. 1981.
151. J. C. Pitt, *Pictures, Images, and Conceptual Change. An Analysis of Wilfrid Sellars' Philosophy of Science*. 1981.
152. R. Hilpinen (ed.), *New Studies in Deontic Logic. Norms, Actions, and the Foundations of Ethics*. 1981.
153. C. Dilworth, *Scientific Progress. A Study Concerning the Nature of the Relation Between Successive Scientific Theories*. 1981.
154. D. W. Smith and R. McIntyre, *Husserl and Intentionality. A Study of Mind, Meaning, and Language*. 1982.
155. R. J. Nelson, *The Logic of Mind*. 1982.
156. J. F. A. K. van Benthem, *The Logic of Time. A Model-Theoretic Investigation into the Varieties of Temporal Ontology, and Temporal Discourse*. 1982.
157. R: Swinburne (ed.), *Space, Time and Causality*. 1982.

158. R. D. Rozenkrantz, *E. T. Jaynes: Papers on Probability, Statistics and Statistical Physics.* 1983.
159. T. Chapman, *Time: A Philosophical Analysis.* 1982.
160. E. N. Zalta, *Abstract Objects. An Introduction to Axiomatic Metaphysics.* 1983.
161. S. Harding and M. B. Hintikka (eds.), *Discovering Reality. Feminist Perspectives on Epistemology, Metaphysics, Methodology, and Philosophy of Science.* 1983.
162. M. A. Stewart (ed.), *Law, Morality and Rights.* 1983.
163. D. Mayr and G. Süssmann (eds.), *Space, Time, and Mechanics. Basic Structure of a Physical Theory.* 1983.
164. D. Gabbay and F. Guenthner (eds.), *Handbook of Philosophical Logic.* Vol. I. 1983.
165. D. Gabbay and F. Guenthner (eds.), *Handbook of Philosophical Logic.* Vol. II. 1984.
166. D. Gabbay and F. Guenthner (eds.), *Handbook of Philosophical Logic.* Vol. III. 1985.
167. D. Gabbay and F. Guenthner (eds.), *Handbook of Philosophical Logic.* Vol. IV, forthcoming.
168. Andrew, J. I. Jones, *Communication and Meaning.* 1983.
169. Melvin Fitting, *Proof Methods for Modal and Intuitionistic Logics.* 1983.
170. Joseph Margolis, *Culture and Cultural Entities.* 1984.
171. Raimo Tuomela, *A Theory of Social Action.* 1984.
172. Jorge J. E. Gracia, Eduardo Rabossi, Enrique Villanueva, and Marcelo Dascal (eds.), *Philosophical Analysis in Latin America.* 1984.
173. Paul Ziff, *Epistemic Analysis. A Coherence Theory of Knowledge.* 1984.
174. Paul Ziff, *Antiaesthetics. An Appreciation of the Cow with the Subtile Nose.* 1984.
175. Wolfgang Balzer, David A. Pearce, and Heinz-Jürgen Schmidt (eds.), *Reduction in Science. Structure, Examples, Philosophical Problems.* 1984.
176. Aleksander Peczenik, Lars Lindahl, and Bert van Roermund (eds.), *Theory of Legal Science. Proceedings of the Conference on Legal Theory and Philosophy of Science, Lund, Sweden, December 11–14, 1983.* 1984.
177. Ilkka Niiniluoto, *Is Science Progressive?* 1984.
178. Binal Matilal and Jaysankar Lal Shaw (eds.), *Exploratory Essays in Current Theories and Classical Indian Theories of Meaning and Reference.* 1985.
179. Peter Kroes, *Time: Its Structure and Role in Physical Theories.* 1985.
180. James H. Fetzer, *Sociobiology and Epistemology,* 1985.
181. L. Haaparanta and J. Hintikka, *Frege Synthesized. Essays on the Philosophical and Foundational Work of Gottlob Frege.* 1986.
182. Michael Detlefsen, *Hilbert's Program. An Essay on Mathematical Instrumentalism.* 1986.
183. James L. Golden and Joseph J. Pilotta (eds.), *Practical Reasoning in Human Affairs. Studies in Honor of Chaim Perelman.* 1986.
184. Henk Zandvoort, *Models of Scientific Development and the Case of Nuclear Magnetic Resonance.* 1986.
185. Ilkka Niiniluoto, *Truthlikeness.* 1987
186. Wolfgang Balzer, C. Ulises Moulines, and Joseph D. Sneed, *An Architectonic for Science.* 1987

GPSR Compliance

The European Union's (EU) General Product Safety Regulation (GPSR) is a set of rules that requires consumer products to be safe and our obligations to ensure this.

If you have any concerns about our products, you can contact us on ProductSafety@springernature.com

In case Publisher is established outside the EU, the EU authorized representative is:

Springer Nature Customer Service Center GmbH
Europaplatz 3
69115 Heidelberg, Germany

Batch number: 10607212

Printed by Printforce, the Netherlands